Tagungsband des 3. Kongresses Montage Handhabung Industrieroboter

Thorsten Schüppstuhl · Kirsten Tracht · Jörg Franke
(Hrsg.)

Tagungsband des 3. Kongresses Montage Handhabung Industrieroboter

Herausgeber
Thorsten Schüppstuhl
Institut für Flugzeug-Produktionstechnik
Technische Universität Hamburg
Hamburg, Deutschland

Jörg Franke
Lehrstuhl für Fertigungsautomatisierung
und Produktionssystematik
Friedrich-Alexander-Universität Erlangen
Erlangen, Deutschland

Kirsten Tracht
Bremer Institut für Strukturmechanik und
Produktionsanlagen
Universität in Bremen (bime)
Bremen, Deutschland

ISBN 978-3-662-56713-5   ISBN 978-3-662-56714-2   (eBook)
https://doi.org/10.1007/978-3-662-56714-2

Die Deutsche Nationalbibliothek verzeichnet diese Publikation in der Deutschen Nationalbibliografie;
detaillierte bibliografische Daten sind im Internet über http://dnb.d-nb.de abrufbar.

Springer Vieweg
© Springer-Verlag GmbH Deutschland, ein Teil von Springer Nature 2018

Gedruckt auf säurefreiem und chlorfrei gebleichtem Papier

Springer Vieweg ist ein Imprint der eingetragenen Gesellschaft Springer-Verlag GmbH, DE und ist ein
Teil von Springer Nature
Die Anschrift der Gesellschaft ist: Heidelberger Platz 3, 14197 Berlin, Germany

Inhalt

Task-based Potential Analysis for Human-Robot Collaboration
within Assembly Systems.. 1
Matthias Linsinger, Martin Sudhoff, Kai Lemmerz, Paul Glogowski und
Bernd Kuhlenkötter

Assembly workshops for acquiring and integrating expert knowledge
into assembly process planning using rapid prototyping model 13
Katharina Gebert, Ann-Kathrin Onken und Kirsten Tracht

Development of a manufacturing cell for fixtureless joining in car
body assembly... 23
Daniel Kupzik, Simon Haas, Dominik Stemler, Jens Pfeifle und Jürgen Fleischer

An Architecture for Intuitive Programming and Robust Execution of
Industrial Robot Programs...................................... 31
Eric M. Orendt und Dominik Henrich

Augmented Reality for programming collaborative robots based on a
physical simulation... 41
Patrick Rückert, Finn Meiners und Kirsten Tracht

Visualizing trajectories for industrial robots from sampling-based
path planning on mobile devices................................. 46
Jan Guhl, Axel Vick, Vojtech Vonasek und Jörg Krüg

Force-controlled Solution for Non-destructive Handling of
Sensitive Objects.. 57
Stefanie Spies, Miguel A V. Portela, Matthias Bartelt, Alfred Hypki,
Benedikt Ebert, Ricardo E. Ramirez und Bernd Kuhlenkötter

Bag Bin-Picking Based on an Adjustable, Sensor-Integrated Suction
Gripper 65
Andreas Blank, Julian Sessner, In Seong Yoo, Maximilian Metzne,
Felix Deichsel, Tobias Diercks, David Eberlein, Dominik Felden,
Alexander Leser und Jörg Franke

A Gripper System Design method for the Handling of Textile Parts   73
Fabian Ballier, Tobias Dmytruk und Jürgen Fleischer

**Automated Additive Manufacturing of Concrete Structures without
Formwork - Concept for Path Planning**   83
Serhat Ibrahim, Alexander Olbrich, Hendrik Lindemann, Roman Gerbers,
Harald Kloft, Klaus Dröder und Annika Raatz

**Sensor-Based Robot Programming for Automated Manufacturing
of High Orthogonal Volume Structures**   93
André Harmel und Alexander Zych

**Finite Element Analysis as a Key Functionality for eRobotics to
Predict the Interdependencies between Robot Control and Structural
Deformation.** 103
Dorit Kaufmann und Jürgen Roßmann

**Semantically enriched spatial modelling of industrial indoor
environments enabling location-based services.** 111
Arne Wendt, Michael Brand und Thorsten Schüppstuhl

**Comparison of practically applicable mathematical descriptions of
orientation and rotation in the three-dimensional Euclidean space** 123
R. Müller, M. Vette und A. Kanso

**"Human-In-The-Loop"- Virtual Commissioning of Human-Robot
Collaboration Systems.** 131
Maximilian Metzner, Jochen Bönig, Andreas Blank, Eike Schäffer und
Jörg Franke

**Simulation-based Verification with Experimentable Digital Twins
in Virtual Testbeds** 139
Ulrich Dahmen und Jürgen Roßmann

**A Flexible Framework to Model and Evaluate Factory Control
Systems in Virtual Testbeds.** 149
Tim Delbrügger und Jürgen Roßmann

**System architecture and conception of a standardized robot
configurator based on microservices** 159
Eike Schäffer, Tobias Pownuk, Joonas Walberer, Andreas Fischer,
Jan-Peter Schulz, Marco Kleinschnitz, Matthias Bartelt, Bernd Kuhlenkötter
und Jörg Franke

**Ontologies as a solution for current challenges of automation of the
pilot phase in the automotive industry.** 167
D. Eilenberger und U. Berger

Reconfiguration Assistance for Cyber-Physical Production Systems 177
André Hengstebeck, André Barthelmey und Jochen Deuse

**Supporting manual assembly through merging live position data
and 3D-CAD data using a worker information system** 187
David Meinel, Florian Ehler, Melanie Lipka und Jörg Franke

Intuitive Assembly Support System Using Augmented Reality 195
Sebastian Blankemeyer, Rolf Wiemann und Annika Raatz

**GroundSim: Animating Human Agents for Validated Workspace
Monitoring** 205
Kim Wölfel, Tobias Werner und Dominik Henrich

**Concept of Distributed Interpolation for Skill-Based Manufacturing
with Real-Time Communication.** 215
Caren Dripke, Ben Schneider, Mihai Dragan, Alois Zoitl und Alexander Verl

**Service-oriented Communication and Control System Architecture
for Dynamically Interconnected Assembly Systems** 223
Sven Jung, Dennis Grunert und Robert H. Schmitt

**Predictive Control for Robot-Assisted Assembly in Motion within
Free Float Flawless Assembly** 231
Christoph Nicksch, Christoph Storm, Felix Bertelsmeier und Robert Schmitt

**Towards soft, adjustable, self-unfoldable and modular robots – an
approach, concept and exemplary application** 241
Robert Weidner, Tobias Meyer und Jens P. Wulfsberg

**Improving Path Accuracy in Varying Pick and Place Processes by
means of Trajectory Optimization using Iterative Learning Control** 251
Daniel Kaczor, Tobias Recker, Svenja Tappe und Tobias Ortmaier

**Path Guiding Support for a Semi-automatic System for Scarfing of
CFRP Structures** 261
Rebecca Rodeck und Thorsten Schüppstuhl

Task-based Potential Analysis for Human-Robot Collaboration within Assembly Systems

Matthias Linsinger, Martin Sudhoff, Kai Lemmerz, Paul Glogowski, and Bernd Kuhlenkötter

Ruhr-University of Bochum, Chair of Production Systems
Universitätsstr. 150, 44801 Bochum, Germany
linsinger@lps.rub.de
WWW home page: http://www.lps.ruhr-uni-bochum.de/

Abstract. The human-robot collaboration (HRC) is an integral part of numerous research activities. The combination of the positive capabilities of human workers and robots ensures a high resource efficiency and productivity. However, especially the planning and implementation of HRC applications requires the consideration of several issues like the economic feasibility, acceptance of employees and in particular technical practicability. Based on the study of existing approaches which consider different technical or economic issues, we present a new method to identify suitable areas of HRC applications within assembly systems and to estimate the optimal allocation of human-related and automated subtasks. Finally, our approach is verified in an assembly line for terminal strips which in turn arises from an industrial use case.

Keywords: Human-robot collaboration, hybrid assembly planning, capability-orientated subtask allocation

1 Introduction

Producing companies are confronted with numerous challenges in the context of globalization. They must react flexibly to a growing range of variants, small batch sizes and shorter product life cycles, without any major production changes. This applies in particular to the cost-intensive assembly. Since neither fully automated nor purely manual assembly systems meet these requirements, hybrid assembly systems are gaining in importance. In this context, human-robot collaboration (HRC) is an integral part of numerous research activities, whereby the positive capabilities of human workers (flexibility, intuition, creativity) and robots (strength, endurance, speed, precision) are combined. This ensures a high resource efficiency and productivity, even in the case of low batch sizes and a high variety in production.

Particular importance is attached to the planning and designing of the collaborative assembly process. [1] This applies especially to small and medium-sized enterprises which cover the entire range from purely manual to fully automated assembly systems. Manual assembly processes can be re-evaluated for their HRC

© Springer-Verlag GmbH Deutschland, ein Teil von Springer Nature 2018
T. Schüppstuhl et al. (Hrsg.), *Tagungsband des 3. Kongresses Montage*
Handhabung Industrieroboter, https://doi.org/10.1007/978-3-662-56714-2_1

potential (e.g. in terms of economic, qualitative or ergonomic benefits) where traditional robotics could not be applied in the past. In order to find the best possible use of the capabilities of human workers and robots in the collaborative assembly process, a goal-oriented allocation of the individual assembly tasks on human workers or robots is an essential aspect. The central question to be clarified is which assembly tasks can be automated and which are carried out manually.

This paper proposes a method for the potential analysis of an HRC system and the resulting capability-oriented allocation of assembly tasks on human workers and robots. Based on a list of criteria of Beumelburg [2], the suitability of human workers and robots is determined for individual tasks within an assembly system for the production of terminal strips. The aim of these investigations is to identify a suitable HRC application within the considered assembly system, since the assembly of the various product variants is at present purely manual.

The paper is structured as follows. First, it describes the existing planning methods of hybrid assembly systems. Thereby, the focus is on the task allocation between the human worker and the robot. As a next step, the paper presents the procedure of the HRC potential identification method. Our work finally presents the industrial application at the Chair of Production Systems (LPS) at Ruhr-University Bochum.

2 Planning of Hybrid Assembly Workstations - Task Allocation between Human Worker and Robot

The implementation of new technologies such as HRC within already established production systems requires the consideration of several issues like the economic feasibility, acceptance of employees and especially technical practicability. These divergent aspects demand a systematic planning. However, many companies and system integrators still lack the necessary expertise in this area [3]. This is one reason why so far only a few workstations with a direct human-robot cooperation or collaboration have been successfully implemented for industrial purposes. If a company strives to transform existing and purely manual workplaces into hybrid HRC systems, both the identification of suitable workplaces and the optimal allocation of the individual assembly tasks between the human workers and robots are the main challenges in planning [4]. Thus, especially the temporal coupling of the human worker and the robot is an essential element of partially automated workplaces [5]. Accordingly, both purely manual and fully automated sub-processes are described separately in the classical sense [6] by determining the time with the help of the *Methods-Time Measurement* (MTM) for the human workers and by using motion simulations for the automation technology which has been used. After this separate approach, both parts are optimized jointly on the basis of precedence graphs and time diagrams. However, the direct HRC or non-separate workspaces of man and machine are not taken into account in this procedure. Nevertheless, there are several existing approaches which consider different technical or economic issues to identify workplaces with high potential for this technology and to estimate the optimal task allocation of man and automation.

The method developed by Beumelburg [2] is known as one of the first approaches to the capability-oriented assembly sequence planning for hybrid workstations. The main focus of this analytical method is the consideration of the assembly process itself, the ergonomics, the component and the provision of parts. Using an evolutionary algorithm, solutions for the allocation of tasks on human workers and robots are developed in the context of automated scheduling. A similar approach, in which parameters and descriptions of their respective task suitability are also assigned to the human worker and the machine, is implemented in [7]. The approach of Takata and Hirano [8] focuses mainly on assembly times and costs. Both suitable functional equipment assignments as well as meaningful automation degrees are optimized for the respective assembly task. However, in the above-mentioned methods [2], [7] and [8], the use of a direct robot assistance is not discussed.

Based on a detailed capability analysis of human workers and robots, Müller et al. [9] have developed a systematic approach in order to enable the capability-oriented allocation of tasks taking into account the required product and process characteristics. The description model used for the definition of assembly tasks is based on the results of Kluge [10]. Accordingly, assembly tasks are characterized by a three-stage model. Müller et al. [9] apply this description model for the derivation of requirements, which are presented as relevant input data for a capability-oriented task allocation on human workers and robots via a pairwise comparison. The approach integrates additional aspects such as the analysis of possible stressful situations of the employees [11], the identification of the employees' acceptance to use the robots, or the recommendation for necessary safety strategies for possible collaborative monitoring steps.

A further method for the capability-oriented assembly sequence planning is presented in [4]. The method outlined here is a three-step process, in which the assembly tasks are described in a break-down principle. The considered level of investigation then refers to the so-called task level. A task is defined here as, for example, the provision or screwing of components. Afterwards, the assignment of invariable tasks takes place, tasks which can be carried out either by the robot or by the human worker and which cannot be distributed. Finally, the variable tasks are defined and assigned. The procedure is based on the approach of Beumelburg [2], since a parametric suitability judgment is applied here as well. Compared to previous methods, the development of a new criteria checklist as well as the differentiation between variably and invariably distributable tasks are the unique selling points.

The approach of Thomas [12] presents a universal evaluation and decision-support system for the human-robot collaboration so that a dynamic task allocation between human workers and automation components is possible. The main goals of the system are to evaluate production variants with three typical degrees of automation (manual, hybrid and full automated assembly), to be able to compare the variants, to precisely consider when to use HRC and to uncover certain exclusion criteria. The criteria developed in [12] are categorized into the five categories of staff, technical conditions, parts, assembly process and production process. In order to make the criteria and categories scalable for companies and thus enable them to define a focus in the evaluation, Thomas has also introduced additional weighting factors. The weighting of each class in terms of the overall rating is based on the number of result-relevant criteria of a class in relation to the relevant criteria of all classes. Each class also contains criteria which exclude a possible production variant when appropriately evaluated. These exclusion criteria include e. g. the qualification of the personnel groups, the gripping and manipulability of the components to be handled or the availability of HRC-capable robots.

However, an integrated approach for identifying a suitable area of application within an assembly system is still missing. Hence, the following method has been developed on the basis of the above-mentioned approaches.

3   Identification of Potential Workstations with HRC

In order to reduce the planning effort, the proceeding for the identification of human-robot collaboration potentials uses a five-step top-down method (see fig. 1). At first the overall assembly line is analysed regarding a fundamental HRC applicability (*step 1*). Afterwards, the individual stations are rated in terms of their HRC potential (*step 2*). This includes a cycle time analysis as well as technical issues. The next step is about a detailed assembly task examination at the assembly station with the highest HRC potential (*step 3*). The MTM procedure is used to analyse the exact motion sequence. This serves to derive the application possibilities and to rate these on the basis of their potential to increase in the productivity and the technical difficulty (*step 4*). The final step provides a recommendation for an HRC application and describes the implementation (*step 5*).

Step 1 - Holistic applicability check of the assembly system. The first step of the approach includes the determination of the fundamental applicability of an HRC implementation within an assembly system. For this purpose, it is necessary to classify various aspects, such as the product structure or the production volume. Among other things, these aspects are important for determining an appropriate automation level. With regard to the classifications by Kluge [10], an HRC application as a hybrid assembly system is well suited for customized products as well as versatile small and medium series production. Furthermore, the high flexibility and mutability of this technology enable shorter product cycles and a high product variety [13, 14].

Step 2 - Cycle times and technical suitability of the assembly stations. In case of a fundamental HRC applicability, the solution space must be curtailed. In order to exploit the assembly station with the highest potential for the HRC application, the stations are analysed regarding the process relevance

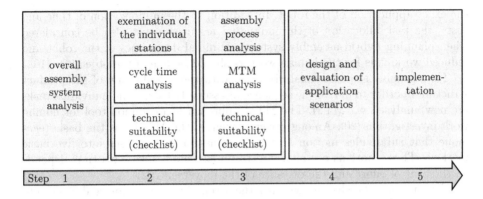

Fig. 1. HRC potential analysis procedure

within the overall system and the technical suitability of the assembly process. The combination of these two aspects can be summarized as the HRC potential degree.

Within a production line, the overall assembly task is divided into several subtasks at each station. Thus, every assembly station has an individual workload that determines its process time. The station with the highest process time dictates the overall cycle time. The main goal of planning assembly lines is to establish a well-balanced process so that every station has a similar workload and cycle time. [15, 16] As a consequence, the process time of a station in relation to the overall process time is a measurable characteristic for its relevance. Hence, the HRC application should be implemented at the station with high workload. On the one hand, it leads to a reduced overall cycle time and, on the other hand, it will be easier to minimize robot downtimes and increase economic efficiency. This is an important aspect for a better economic feasibility [3].

The *technical suitability* of an HRC application depends on many factors. The most important attributes are component characteristics and process properties. In addition to that, the part supplying process is also an essential issue of the assembly process. Furthermore, standardized safety requirements have to be considered as well as ergonomic issues. These aspects are included in a criteria checklist which is based on [2] and [17]. In addition to these approaches, the methods from Kluge [10], Bullinger [18] and Ross [19] are also considered. If a characteristic of a criterion applies, it must be decided whether a manual, a collaborative or a fully automated assembly system can best perform the process in terms of process time, quality and investment. According to their individual characteristics, each station finally receives a value for its HRC suitability. Both the HRC suitability and the process relevance must be equally considered. In order to ensure a high HRC potential, it is essential to provide a high workload as well as sufficient technical practicability.

Step 3 – Assembly sequence analysis. The last evaluation step determines the assembly station with the highest HRC potential. In this context, it defines the exact application of the robot. In addition to the minimization of time and costs, the task allocation of the individual resources must also be considered when planning hybrid assembly systems. Minimal downtimes of the robot and reduced workloads for the human workers should be one of the objectives [16].

The assembly sequence analysis starts with the examination of the product structure in order to derive the sequence of assembly tasks. The individual tasks are now analysed via MTM. The MTM approach is a common tool for human work investigation [20]. Among several variants, the MTM-1 is the basic technique that subdivides motion sequences of assembly processes into five basic motions. These motions are allocated to an expected time. This time depends on the control effort and the distance of the movement. [21]

Now it has to be checked whether the assembly process can be divided into sections which should be operated either automatically or manually [6]. Especially in the HRC context, it is necessary to identify potential collaborative sections (hybrid assembly). For this purpose, a technical suitability check as described in step 2 is carried out for each relevant subtask. In addition to that, the task sections are examined with regard to their value creation ratio.

Step 4 and 5 – Design of the application scenarios. Based on the previous analysis, an appropriate field of application is defined as part of this step. The application scenarios are derived and assessed in terms of the productivity increase potential, the technical feasibility as well as ergonomic aspects. Finally, this step describes the most suitable application and its implementation in the existing assembly system.

4   Industrial Application

The method has been verified in an assembly line for terminal strips as they are installed in switch cabinets. The assembly line is operated by the LPS in cooperation with Phoenix Contact GmbH & Co. KG. The motivation to run an assembly line production for industrial customers at LPS is to develop and verify new assembly processes and technologies under industrially relevant conditions.

4.1   Assembly Line

The assembly of switch cabinets and especially terminal strips is characterized by predominantly manual assembly operations as the product variance and number of different parts is extremely high. Hence, automation has rarely been implemented in terminal strip assembly processes as the technology has not been able to cope with the high product variance. [22]

The u-shaped assembly line at LPS consists of six stations. After cutting the strips in individually required lengths on a pre-assembly station, terminals are mounted onto the strip at *station 1*. Depending on the variant, spacing may be required between terminals. End terminals are designed to hold sections of terminals in place on the strip and prevent them from slipping.

A labelling of terminals is carried out at *station 2*. Labels are required to identify terminals according to the circuit diagram of the switch cabinet, e. g. to conduct the final wiring. *Station 3* is provided for the circuit bridge and plug assembly. Circuit bridges and plugs are pressed into the terminals in order to create an electrical connection between different terminals. In a quality inspection at *station 4*, no parts are assembled but the product is inspected against deviations. For some customers, the terminal strip is equipped with pre-assembled cabling. Cabling is conducted at *station 5*. At *station 6*, the finished product is packed and prepared for shipping.

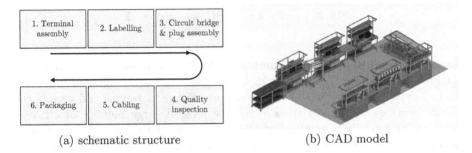

(a) schematic structure   (b) CAD model

Fig. 2. Terminal strip assembly line at LPS

4.2   Evaluation and Results

Step 1 - Holistic applicability check of the assembly system. Within the present assembly system, mainly small series are produced. Furthermore, the product is simply structured and can be configured by the customer from a large variety of standardised components. Accordingly, it is a standardized product with customer chosen variations. As described in section 4.1, the assembly process is carried out completely manually. In summary, the characteristics of the assembly system correspond to the conditions of hybrid assembly systems mentioned in chapter 3.

Step 2 - Process relevance and technical suitability of the single assembly stations. For the identification of the assembly station with the highest HRC potential, a cycle time analysis has been performed. As shown in figure 3(a), the process times at the terminal assembly and labelling stations are the highest.

As a conclusion of the technical suitability check, it can be stated that most stations of this assembly system have a fundamental HRC applicability. However, there are differences in the technical realization. Thus, the need for parts' separation or sensory systems will cause additional costs. The process reliability is also more difficult to achieve. Therefore, the stations for terminal and bridge assembly can be identified as the most appropriate HRC application fields. By taking into consideration the HRC suitability degree (see fig. 3(a)), it is the terminal assembly station which is finally chosen for our purposes.

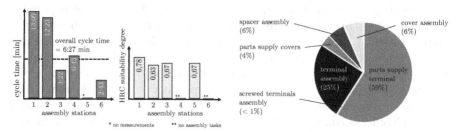

(a) cycle times and HRC suitability degree (b) component group analysis for station 1

Fig. 3. Results of the HRC potential check

Step 3 - Assembly sequence analysis. The purpose of the assembly sequence analysis is to find out the subtasks that correspond to the strengths of the robot. Accordingly, the robot should rather perform the monotonous tasks while the human workers performs the more volatile jobs. A large number of different components must be assembled at station 1 in order to generate a customer-specific product. For the MTM analysis, these components have been classified into groups. Each component group has a different assembly sequence. These sequences have been subdivided into the MTM basic processes and, consequently, analysed with regard to their duration and repetition rate.

It has been shown that the terminals without cover and screw connection are by far the most common type of components. The importance of the two subtasks (parts supply terminal and terminal assembly) for assembling this type of terminals are correspondingly high. This correlation is illustrated in figure 3(b). At 59 % and 25 %, the time shares of these two assembly subtasks account for by far the largest proportion of the total cycle time. In contrast to this, the time portions for the cover assembly and terminals with a required screw are below 10 %.

Step 4 – Design of the application scenarios. Therefore, against the background of the previous analysis, both the subtasks for assembling a screw-fastened terminal as well as the cover assembly should be carried out by the human worker. On the one hand, this has the reason that only about a sixth of the terminals of an average terminal strip require a cover. On the other hand, the cover dimensions are inappropriate, since the outer surfaces are very small and the clamp assembly requires a certain dexterity. Taking into account these evaluations, three scenarios for automation or partial automation have been identified (see fig. 4).

1. Robot performs terminal part supply, human worker mounts the terminal
2. Human worker performs terminal part supply, robot mounts the terminal
3. Robot performs terminal part supply and mounts the terminal

In each scenario, the human worker and the robot act simultaneously and complement each other within the process of assembling the terminal block. The advantage of the *first scenario* is the comparatively simple technical implementation. The parts can be provided at defined points and the handover point does not require an exact definition. The disadvantage, however, is the high ergonomic strain on the employee due to the monotonous task and forces to be applied. These process forces are not required in the *second scenario*, as the robot is responsible for mounting. This means an increased technical difficulty, as the assembly location always changes and high tolerances have to be observed. These technical difficulties also arise in the *third scenario*. Here, however, the transfer point is not necessary. In addition, the ergonomic strain on the human worker is significantly reduced. The human worker only has to carry out upstream and downstream subtasks such as assembling the covers. According to these aspects, the third HRC scenario is assumed to provide the highest process benefit.

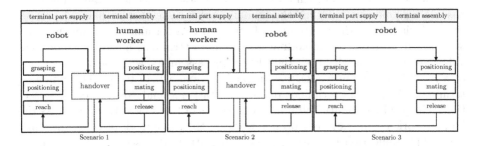

Fig. 4. HRC application scenarios

4.3   HRC Application

Using the results of the potential analysis, an HRC application has been detailed based on the third scenario where the robot grasps, picks up and mounts the end terminals onto the strip. For this purpose, station 1 has been extended to the left for a robot-based end terminal pre-assembly (see fig. 5). Before the operation starts, the strip is placed onto a fixture by the human worker. After that, the robot mounts the end terminals with a 2-finger gripper onto the strip. The human worker takes over the workpiece to the right part of the station and mounts the remaining terminals between the already assembled end terminals.

This application shows potential improvement of the current, purely manual assembly process in terms of *process efficiency, ergonomics* as well as *product quality*. Comparing the required steps to set the first end terminal on a terminal strip, the robot requires less than half as many steps as the human worker. For example, taking the MTM analysis into account, checking, grasping, inserting and returning spacers takes almost 4 seconds for the human worker. The robot does not need to set spacers at all because distances between terminals are integrated in the robot's path. As experimental measurements at LPS have shown,

Fig. 5. HRC concept for terminal assembly

end terminals require a mounting force up to 240 N. Hence, the end terminal assembly leaves a high burden on the human worker. Therefore, the application ensures a significant relief for the human worker. The current manual process shows deviations in the positioning of terminals. As terminals are built up from left to right, tolerances are adding up the more terminals have been mounted. When end terminals with their tight fit are pre-assembled with the high accuracy of a robot, the human worker has defined sections where to set the remaining terminals. As the end terminals define the start and end position of these sections, the product quality will improve in terms of less staged tolerances. As the human worker and the robot do not work on the same work piece at the same time, this application is considered as a coexisting HRC scenario (for example see [2] for definitions). As mounting forces of 240 N extend the allowed forces for direct human-robot collaboration (see [23]), the safety concept must preclude the human worker to grasp between the robot gripper and the work object during the assembly process (e. g. by using workspace monitoring with sensors and cameras).

5 Summary and Outlook

This paper describes a five-step top-down method to identify potentials for human-robot collaboration in manually based assembly systems. This integrated approach takes manual process cycle times, technical suitability checklists as well as a detailed MTM-based assembly process analysis into account. In contrast to previous potential methods, single assembly tasks can be assigned to the human worker or the robot. Based on that, different HRC scenarios can be generated and compared in terms of technical suitability, degree of automation, human-robot interaction intensity, ergonomics and task allocation. Depending on the type and intensity of human-robot interaction, a rough safety concept can be derived as well for each scenario.

The evaluation in the assembly system for terminal strips at LPS has proven that a reasonable direction for the HRC potential and application can be given with this method without the need to integrate experts. However, as HRC is still an emerging technology in industry, it is indeed recommended to involve experts and representatives from the worker's association when a chosen HRC scenario needs to be detailed for implementation: Safety approval and CE compliance labelling by authorities is crucial to implement HRC in industry and cannot be guaranteed in the early planning phase when this method is applied.

The HRC application will have been implemented at LPS by June 2018. The results can be evaluated immediately for its industrial applicability as the assembly system is regularly loaded with orders from industry and uses identical processes.

References

1. Glogowski, P.; Lemmerz, K.; Schulte, L.; Barthelmey, A.; Hypki, A.; Kuhlenkötter, B.; Deuse, J.: Task-based Simulation Tool for Human-Robot Collaboration within Assembly Systems. In: Tagungsband des 2. Kongresses Montage Handhabung Industrieroboter. Ed. by Schüppstuhl, T.; Franke, J.; Tracht, K. Springer Berlin Heidelberg (2017), pp. 155-163.
2. Beumelburg, K.: Fähigkeitsorientierte Montageablaufplanung in der direkten Mensch-Roboter-Kooperation. PhD thesis, University Stuttgart, Jost-Jetter Verlag (2005).
3. Bauer, W.; Scholtz, O.; Braun, M.; Rally, P.; Bender, M.: Leichtbauroboter in der manuellen Montage - Einfach einfach anfangen. Erste Erfahrungen von Anwenderunternehmen (2016), pp. 1–63.
4. Ranz, F.; Hummel, V.; Sihn, W.: Capability-based Task Allocation in Human-robot Collaboration. In: Procedia Manufacturing 9 (2017), pp. 182–189.
5. Lotter, B.; Wiendahl, H.-P.: Montage in der industriellen Produktion. Ein Handbuch für die Praxis. Berlin, New York: Springer (2006).
6. Pfrang, W.: Rechnergestützte und graphische Planung manueller und teilautomatisierter Arbeitsplätze. PhD thesis, TU München (1990).
7. Zülch, G.; Becker, M.: A Simulation-supported Approach for Man-Machine Configuration in Manufacturing. In: International Journal of Production Economics 125 (2010) 1, pp. 41–51.
8. Takata, S.; Hirano, T.: Human and Robot Allocation Method for Hybrid Assembly Systems. In: CIRP Annals 60 (2011) 1, pp. 9–12.
9. Müller, R.; Vette, M.; Mailahn, O.: Process-oriented Task Assignment for Assembly Processes with Human-Robot Interaction. In: Procedia CIRP 44 (2016), pp. 210–215.
10. Kluge, S.: Methodik zur fähigkeitsbasierten Planung modularer Montagesysteme. PhD thesis, University Stuttgart (2011).
11. Spillner, R.: Einsatz und Planung von Roboterassistenz zur Berücksichtigung von Leistungswandlungen in der Produktion. PhD thesis, TU München (2014).
12. Thomas, C.: Entwicklung einer Bewertungssystematik für die Mensch-Roboter-Kollaboration. PhD thesis, Ruhr-University Bochum (2017).
13. Thiemermann, S.: Direkte Mensch-Roboter-Kooperation in der Kleinteilemontage mit einem SCARA-Roboter. PhD thesis, University Stuttgart (2004).
14. Krüger, J.; Lien, T. K.; Verl, A.: Cooperation of Human and Machines in Assembly Lines. In: CIRP Annals-Manufacturing Technology 58 (2009) 2, pp. 628–646.

15. Deuse, J.; Busch, F.: Zeitwirtschaft in der Montage. In: Lotter, B.; Wiendahl, H.-P.: Montage in der industriellen Produktion. Berlin, Heidelberg: Springer Berlin Heidelberg (2012), pp. 79–107.
16. Chen, F.; Sekiyama, K.; Cannella, F.; Fukuda, T.: Optimal Subtask Allocation for Human and Robot Collaboration within Hybrid Assembly System. In: IEEE Transactions on Automation Science and Engineering 11 (2014) 4, pp. 1065–1075.
17. Naumann, M.: Montageprozesse wirtschaftlicher gestalten. In: Maschinenmarkt, Das Industriemagazin (2016) 23, pp. 48–51.
18. Bullinger, H.-J.: Ergonomie. Produkt- und Arbeitsplatzgestaltung. Technologiemanagement - Wettbewerbsfähige Technologieentwicklung und Arbeitsgestaltung. Wiesbaden, Vieweg+Teubner Verlag (1994), p. 1.
19. Ross, P.: Bestimmung des wirtschaftlichen Automatisierungsgrades von Montageprozessen in der frühen Phase der Montageplanung. PhD thesis, TU München (2002).
20. Reichenbach, M.: Entwicklung einer Planungsumgebung für Montageaufgaben in der wandlungsfähigen Fabrik, dargestellt am Beispiel des impedanzgeregelten Leichtbauroboters. PhD thesis, BTU Cottbus (2010).
21. Bokranz, R.; Landau, K.: Handbuch Industrial Engineering. Produktivitätsmanagement mit MTM. Stuttgart: Schäffer-Poeschel (2012).
22. Großmann, C.; Graeser, O.; Schreiber, A.: ClipX: Auf dem Weg zur Industrialisierung des Schaltschrankbaus. In: Vogel-Heuser, B.; Bauernhansl, T.; ten Hompel, M.: Handbuch Industrie 4.0 Bd. 2. Springer (2017), pp. 169–187.
23. DIN ISO/TS 15066:2016: Roboter und Robotikgeräte – Kollaborierende Roboter. DIN Deutsches Institut für Normung e. V. Berlin: Beuth Verlag GmbH (2017).

Assembly workshops for acquiring and integrating expert knowledge into assembly process planning using rapid prototyping model

Katharina Gebert[1,a], Ann-Kathrin Onken[1,b], Kirsten Tracht[1][0000-0001-9740-3962]

[1] University of Bremen, Bremen Institute for Mechanical Engineering, Badgasteiner Straße 1, 28359 Bremen, Germany
[a]gebert@bime.de, [b]onken@bime.de

Abstract. For the integration of experts' and non-experts' knowledge from the assembly planning into the product development process, assembly workshops are used to acquire the knowledge and to transfer it back to the product development. A workshop was conducted using a 3D printed model of a large-volume assembly group from the aviation industry. For a repeatable implementation of the assembly workshops with a changeable assembly sequence, individual components and connecting segments are structurally simplified, so that a non-destructive assembly and disassembly can be ensured. By using questioning and observation methods the expertise of assemblers and laymen are documented. The findings of the workshops from the laymen and expert groups are evaluated to show the ability of the probands for transferring the assembly structure. For the laymen, the simplification of the components mainly influences this ability. The workshop provides a basic understanding of what experiences or learnings, generated during the assembly, can be considered in order to plan assembly processes. The prerequisite is that the participants can recognise the restrictions and the assembly effort of the real assembly, while using the simplified 3D printed model.

Keywords: Rapid Prototyping, Planning, Assembly Workshop

1 Rapid prototyping in assembly process planning

The increasing demand of companies for product development processes leads to fundamental changes in development concepts. Therefore, it is important to perform quickly and cost-effectively on market, while achieving an efficient result for aesthetics, safety and ergonomics. [1] One industry-established process for validating a product design during the product implementation process is Rapid Prototyping (RP). This process involves the preparation of a physical model by using additive manufacturing, which is based on computer aided design (CAD) data of the product. [2] The RP is often used to make design and geometry adjustments during the early product design stage without time- or cost-intensive investments for preliminary series productions. In the context of product development, the RP is used for examination and learning units in the handling of the product, for assembly preparation or for ergonomic tests. [1, 3] Especially for complex products physical prototyping is still required as an established method in the industry. [4] Gallegos-Nieto et al. evaluate the effectiveness of virtual assembly training in comparison with the traditional assembly training approach using a rapid prototyping model. It becomes obvious, that the real assembly process leads to a faster learning process than in virtual assembly, because learning and cognition is benefited by the realistic haptic interaction. [5] As Ahmad et al. sum up in their research work, RP models offers many benefits as an instrument in assembly training, like physical touch, actual tolerances as well as in assembly sequence planning. [6]

This paper shows how a rapid prototyping model can be used in an assembly workshop, in particular how to gain expertise in assembly planning in order to use it for assembly-friendly product planning. In addition, assembly approaches of laymen can also indicate new solutions for redesigning assembly processes in its structure and sequence. New approaches derived from the obtained information, are used to design the manufacturing and assembly. This case study presents the implementation of the RP model, adapted from a large-volume assembly group from aviation industry, in an assembly workshop. The findings of this paper emphasize what turns out to be useful for assembly process planning. Depending on the assembly expertise, differences are expected in the use of tools and equipment of the participants, in the order of assembly, as also in the kind of challenges they encounter during assembly. Probands with assembly experience could benefit from their own experience of similar products. Expected is a shorter assembly time or a simplified assembly order compared to the laymen group.

2 Conception and planning of the assembly workshop

Considering the assembly process planning in a product development process can significantly influence the product design. However, an integration of the final assembly process for a non-series product is difficult, because neither a structural plan nor the scheduling are available yet. In addition, the implementation of assembly processes is characterized by a repetition of investigations. At the same time, new assembly activities and processes still need to be learned, so that assembly experience has to be established. For this purpose, the first assembly runs performing with a printed model allow to achieve initial knowledge about assembly capability at an early stage of product development.

2.1 3D printed model

The components of the assembly group are designed and modelled by using CAD software. Figure1 shows the 3D printed model of the assembled group. The design and product architecture of the printed 3D model is adapted to the original assembly object, but simplified regarding its shape. In this regard, the number of components is limited and geometry is reduced on its basic geometry. Prerequisite was to design the individual components in such a way that, despite the reduction of the size, can still be gripped and assembled in manual assembly. Unlike the real assembly group, the joints are structurally simplified and conceptually designed for a possible disassembly. In a test run, one person assembled the printed model, which consists of 91 components except screws, in approx. 22 assembly steps.

Fig. 1. 3D printed model of large-volume assembly group with simplified joints

2.2 Assembly workshop observation method

The workshop is based on design tools and methods of data collection from work analysis. Questioning and observation methods are used to monitor and document the information generated in the workshop. The advantages of combining these methods are that they complement each other in many aspects and provide inaccessible information. The examination with a questionnaire is used to investigate activities, which are hard to observe, for example the scope of information acceptance and the processing of the probands. [7] The survey method is suitable for documenting subjective judgments, such as an experience-based load in application-related processes. The implementation of an observation method prevents the loss of internalized, highly automated operations, which are difficult to put into words. [8] Observational methods primarily serve to capture the activities and behaviour objectively, including operations unconscious to the test person itself. [9]

2.3 Assembly workshop planning

Five assembly workshops, which varied in terms of assembly structure planning and workplace layout, were conducted, as shown in the experimental Matrix in Figure 2. For the two groups of assembly experts and the three groups of laymen, specific workshop variants were selected and two variants were compared. Because of the variation in workplace layout, it was compared to which extends the instructions for structured and disorganised workplace layout influences the workflow and the group dynamics.

An advantage of a structured workplace layout is the enabling of natural movements and the reduction of unnecessary stress during assembly. The results of an experimental study on the effect of spatial arrangement of assembly workplaces of Kothiyal and Kayis prove that it has a significant effect on work-cycle time of a manual assembly task. The employee tires faster, which reduces his/her performance significantly. The ergonomic design of the workplace for assembly tasks helps to improve the efficiency of the operator and includes as example the correct posture and environmental factors such as lighting and noise. [10] The arrangement of the components provided is based on the workspace design of Sanders and McCormick, who defined a recommended working area for manual assembly. The left and right working area, which the hands can reach easily by spreading arms, are divided in normal and maximum area. If these areas are maintained during the manual assembly, the stress can be reduced and the productivity can be increased. [11]

The workshop variations with optional and mandatory assembly structure planning demonstrate the comparison between intuitive and guided assembly planning of the probands. These have a decisive influence on the success of the assembly structure and the assembly effort. One workshop unit consists of a group of two experts or two laymen. The workshop process was documented by the workshop leader in writing and was also recorded with a camera. This information is used for evaluation and reflection. In the workshops, the usage of 3D printed components for developing and returning of assembly knowledge in the assembly planning process was evaluated. In order to realize the large-volume assembly group as a 3D model, the number and geometry of the components had to be reduced. Based on this simplification, the assembly operations adapt the essential steps of the real assembly process.

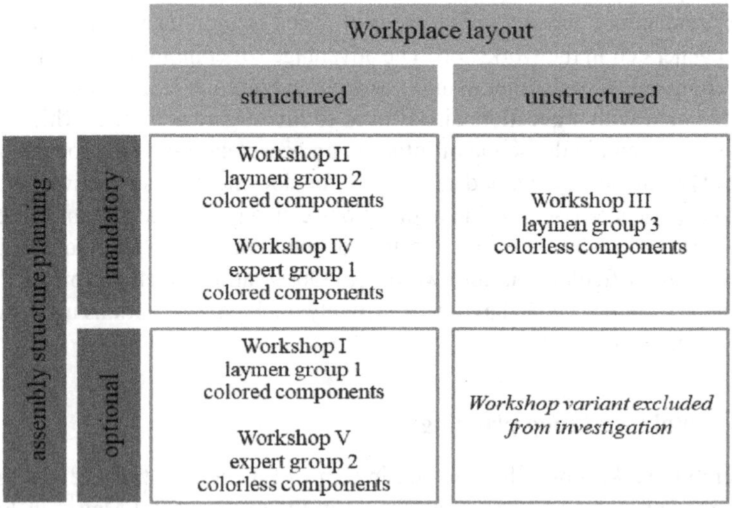

		Workplace layout	
		structured	unstructured
assembly structure planning	mandatory	Workshop II laymen group 2 colored components Workshop IV expert group 1 colored components	Workshop III laymen group 3 colorless components
	optional	Workshop I laymen group 1 colored components Workshop V expert group 2 colorless components	*Workshop variant excluded from investigation*

Fig. 2. Experimental matrix of assembly workshops with variation of workplace design and assembly structure planning.

The part of the movability of the components was retained, since this property is essential for the characteristic for the assembly group and makes the assembly particularly difficult. An important aspect of the assembly process is the joining direction, which was aligned to the real product. Since the joining direction predetermines the assembly operation, the screw joints could not be simplified to plug connection, thus the connecting specific tool could be retained. The component joints were structurally simplified, while a non-destructive assembly and disassembly is retained and the need of using assembly tools is maintained. An essential question that should be answered by the experiment is whether the participants can recognise the assembly effort of the real assembly despite these restrictions and the simplification of the 3D printed model.

3 Integrating 3D printed model in assembly workshop

For the workshops with the assembly experts, it is important that the probands already own previous knowledge and experience with large-volume assembly groups. For laymen it has to be considered that the probands have no experiences with similar assembly groups, so that they are not restricted or influenced by certain experience patterns. At the beginning of the workshops, the probands were introduced to the structure of the assembly workshop. For the implementation of the workshop, writing material, assembly tools as well as an assembled 3D printed model for the orientation, are provided. The technical drawings or CAD data of the original assembly object are deliberately hold back from the test persons in order to cover information about the real dimension and product characteristics, like material, weight, or function. The dimension should not influence the probands in their thoughts or approaches, since the handling with the model should be intuitively. In the first workshop conducted with laymen, the workplace was structured for the probands (Figure 3). The screws and joining parts have already been separated and assigned to the respective joints. The components' colors and number of parts correspond to the pre-assembled model, which could be used as example.

Fig. 3. Structured workplace with assembled 3D printed model provided as an example

The first group with laymen was not instructed to prepare an assembly structure planning, so that the laymen decided whether a work plan would be necessary at the beginning of the assembly. The assembly process was self-organized and without intervention of the workshop leaders. There was no time limit for the respective planning and working phases. After completing the assembly, the probands had to fill out the questionnaire for information regarding the selected assembly order, their challenges during assembly, and the selection of assembly tools. The documents exhibit unconventional approaches, which can be used as potential improvements for assembly planning. A second questionnaire, which is given to the probands after an explanation of actual dimensions, weight and number of components, opens the opportunity to restructure the assembly order. Inquiring ideas for supporting the worker by usage of different assembly tools could lead towards innovative solutions. Due to the simplified and abstracted tasks, new and unconventional knowledge was acquired. New ideas were focused on the assembly operations, for example the spatial orientation of component in joining processes of large components. In the second workshop with the laymen group 2, a structured workplace layout was arranged and the task of creating an assembly structure was given. The third workshop with laymen group 3 varied in terms of workplace layout. Thus, the components were not arranged and had to be structured by the probands autonomously. As in the other workshops before, the probands could rework their considerations after the information about dimension and weight of the components was given. Workshop I and II were repeated with expert group 1 and expert group 2. Furthermore, expert group 2 got the task to use an uncoloured assembly group of the 3D printed model.

4   Findings and results from assembly workshops

An observational method during the workshops was used to detect intuitive operations of the probands during the assembly. Videos of the workshops were recorded with the intention of receiving additional information about the approach and execution of the assembly operations, including the combination of verbal statements and non-verbal operations. These are for example unconscious and experience-oriented decisions, which result in interactions within the group.

4.1   Intuitive assembly operations of laymen

Ergonomical aspects were gained for the assembly process planning from the observation of the intuitive assembly operations during the workshop I, II and III with the laymen groups. The laymen were neither given the handling with the components, nor could they have recourse to their own experience, for positioning the components in order to assemble them as quickly and easily as possible. Accordingly, the components were assembled intuitively. However, this only works if installation-related aspects, such as weight, function, or stability of the assembly are excluded. The simplified system offers a focus on an easy and comfortable assembly position for the assembly operator. The video recordings provide a basic understanding for alternative options for designing the assembly process. The intuitive handling, which was observed, acquires inspirations and impulse for new approaches for assembly process planning for the implementation of new products. A new idea to improve assembly is the arrangement of component groups in a pre-assembly, based on the self-structured assembly order of the laymen.

The influence of a structured workplace layout on the assembly process became obvious from the comparison of laymen groups 1 and 2 to laymen group 3. Complications at the beginning of assembly planning became apparent since the laymen group 3 needed more time for the allocation of the components. The consequence of the disordered assembly workplace was a repetitive restructuring of the assembly sequence, which resulted in unnecessary movement of the assembly object.

4.2 Acquiring knowledge from experiences

The results from the workshops IV and V with the assembly experts show that experiences from the daily work with assembly groups provide essential advantage for assembly planning operations. The assembly planning of the experts was oriented on the assembled 3D printed model. Both expert groups recognized the assembly challenges of the real assembly group, despite the simplification of the model. Whereas, the influence of the colour of the components becomes obvious by observing expert group 1, who oriented on the colours when pre-sorting the assembly group. The self-organised assembly sequence planning of both expert groups show that the experts were prejudiced by their own assembly experiences, therefore the components were assembled according to the conventional assembly process.

4.3 Ideas for assembly devices

The answers of the questionnaires from laymen and expert groups show that in most cases the requirement of assembly tools was recognized. In addition, the use of assembly devices could be assigned to specific assembly situations by all probands, for example in the case of joining the fuselage with the wing or for fine adjustment of the screws. In case that the laymen gave no conscious ideas or suggestions for the use of assembly devices, the video analysis was used additionally for analyzing the assembly processes. Especially joining operations of large 3D printed components show that assembly tools or clamping devices would have decreased the assembly effort. For instance, all probands had to deposit the assembly group repeatedly to search for further components or were forced to change their grip to fix the unstable assembly group (Figure 4).

The probands' handling of the components during the assembly can be transferred into design approaches and positioning of assembly grippers, which increase the assembly access for the operator. In addition, the results from the evaluation of the questionnaire of all laymen groups show that the main challenge for the laymen was the lack of information regarding the dimensions of the real assembly group. Even after the explanation of the real component size and weight, the laymen could not transfer and integrate these information into the considerations of the assembly structure process. The need for assembly devices has also been recognized by the expert groups and even specified regarding to their respective assembly situation. The task sharing started at a very early stage of the planning phase, because arising challenges during the assembly were recognized earlier, thus they could be prevented for the assembly of the printed model. The non-verbal knowledge was achieved by observing the interaction of experts with the 3D printed components and enables the transfer of assembly experiences and intuitions to the assembly planning for the real product group. Thus, working with the simplified model shortens the implementation time of assembly planning processes, because the repetition of investigations are moved in the assembly workshops forward.

Fig. 4. Handling with the 3D printed model during assembly workshop

5   Conclusion and Outlook

This paper illustrates the advantages of the use of 3D printed models in assembly workshops. RP models can be used in assembly workshops to acquire inspirations and impulses for new approaches for assembly process planning during the implementation of a new product. From the observations of the workshops, measures can be derived to offer un-conventional suggestions for assembly process planning. One example is the self-organized assembly structure, which can be used for optimizing the assembly sequence. Especially for the group of laymen it is important that the probands are not influenced by thought patterns or conventional assembly structures. In most cases, the components were divided to assembly sub-groups and were allocated to the group members autonomously. This configuration of component groups can be adapted as an intuitively generated assembly order. The intuitive working with the model can also be used to design an ergonomic and efficient assembly process. Concerning the assembly devices, however, it remains unclear if all challenges of assembly processes can be represented by the RP model adequately. Due to the design simplification, laymen could not recognise the assembly specific requirements completely. The high safety requirements cannot be taken into account sufficiently, because of the high simplification of components and joints. The use of prototypes provides more reliable information, because in contrast to the 3D printed Components, these physical models are almost identical to the original components. However, if the prototype components are not available, aspects regarding the influence of weight, material and dimensions of the assembly objects cannot be identified adequately. One possibility to observe an extension of assembly effort is the use of a derivative product. This product can be used at an early stage of product development to perform assembly studies. It considers all assembly-related aspects, but excludes geometry details and product function. [12] The appropriate degree of functional or structural analogy and abstraction has to be indicated, depending on the case of use of the derivative product. An application-specific design of derivative products will be conducted in future re-search. Based on this results, new assembly workshops will be realized.

Acknowledgments

The results presented in this paper were developed in the research project "Next.Move - Next Generation of Moveables" funded under the programme Luftfahrtforschungs programm LuFoV-2 by Federal Ministry of Economics and Technology (BMWi).

References

1. Chua C. K., Leong K. F., Lim C.S.,: Rapid Prototyping - Second edition, World Scientific Publishing Co. Pte. Ltd., Singapur. pp. 1-33, 2003.
2. Dheeraj N., Mahaveer M. X.: Using Rapid Prototyping Technology In Mechanical Scale Models. International journal of engineering research and applications, pp.215-219, 2003.
3. Yagnik, P. D.: Fused deposition modeling - A rapid prototyping technique for product cycle time reduction cost effectively in aerospace applications, IOSR Journal of Mechanical and Civil Engineering, pp. 62-68, 2014.
4. Maropoulos, P.G., Ceglarek, D.: Design verification and validation in product lifecycle, CIRP Annals Manufacturing Technology, 59(2), pp.740-759, 2010.
5. Gallegos-Nieto, E., Medellin-Castillo, H. I., Gonzalez-Badillo, G., Lim, T., Ritchie, J.: The analysis and evaluation of the influence of haptic-enabled virtual assembly training on real assembly performance, International Journal of Advanced Manufacturing Technology. 89,1, pp.581-598, 2017.
6. Ahamad A., Darmoul S., Ameen W., Abidi M., AL-Ahmari A.M.: Rapid Prototyping For Assembly Training And Validation, In 15th IFAC Symposium on Information Control Problems in Manufacturing, IFAC Papers On Line 2015 48(3), pp.412-417, 2015.
7. Dunckel H., Zapf D., Udris, I.: Methoden betrieblicher Stressanalyse. In S. Greif, E. Bamberg, N. Semmer (ed.), Psychischer Stress am Arbeitsplatz, Gottingen: Hogrefe, pp.29-45, 1991.
8. Nerdinger F.W., Blickle G., Schaper N.: Arbeits- und Organisationspsychologie, Springer-Verlag Berlin Heidelberg 2008, pp.353-380, 2008.
9. Dunckel, H.: Psychologische Arbeitsanalyse: Verfahrensüberblick und Auswahlkriterien. In H. Dunckel (ed.), Handbuch psychologischer Arbeitsanalyseverfahren, Zurich: vdf. pp.9-30, 1999.
10. Kothiyal K.P., Kayis B.: Workplace Design for Manual Assembly Tasks: Effect of Spatial Arrangement on Work-Cycle Time, International Journal of Occupational Safety And Ergonomics, The University of New South Wales, VOL 1, NO. 2, pp.136-143, 1995.
11. Sanders, M. & McCormick, E.: Human Factors in Engineering and Design, Seventh Edition. Addison-Wesley Publishing, Massachusetts, pp.413-455, 1993.
12. Bader A., Gebert K., Hogreve S., Tracht K.: Derivative Products Supporting Product Development and Design for Assembly, 6th International Conference on Through-life Engineering Services, TESConf 2017, Bremen, Germany, 2017.

Development of a manufacturing cell for fixtureless joining in car body assembly

Daniel Kupzik[1,a],Simon Haas[1,b],Dominik Stemler[1,c],Jens Pfeifle[1,d],Jürgen Fleischer[1,e]

[1] Karlsruhe Institute of Technology (KIT), wbk Institute of Production Science,
Kaiserstrasse 12, 76131 Karlsruhe, Germany
[a]daniel.kupzik@kit.edu,[b]simon.haas2@student.kit.edu,[c]dominik.stemler@student.kit.edu,[d]jens.pfeifle@student.kit.edu,[e]juergen.fleischer@kit.edu

Abstract. The integration of electric drivetrain components into the chassis of automobiles dramatically increases the variety of car body designs and calls for changes to the production processes. One approach is to replace the typical assembly line manufacturing process with a workshop production system, as has already happened in machining, where transfer lines have been replaced by flexible manufacturing systems. One requirement for this shift is the development of a flexible joining cell which can perform different stages of joining operations on different variations of a vehicle. One step in the development of such a cell is the replacement of mechanical positioning elements, used to define the parts' relative positions before joining. In this work, the positioning of the components is done using a camera-based measuring system and a control loop to reorient the robots holding the parts. This paper describes how the system is developed and implemented in a demonstration cell.

Keywords: car body assembly, fixtureless joining, optical guidance

1 Introduction

1.1 Market development

The German automotive industry already offers a relatively broad portfolio of electrified vehicle. In addition to plug-in hybrids (PHEVs), the market is also in full swing on purely electrically powered vehicles based on batteries (BEV) or fuel cells (FCEV). The use of these vehicles in municipal or commercial fleets is often profitable and can be an enabler for the wider spread of electric vehicles. The economic production of these electrified vehicles (derivatives) is the decisive investment prerequisite for the ramp-up of the market, especially in the face of the challenge of highly fluctuating and uncertain market forecasts. In addition to production costs, delivery times for vehicles are also an important competitive differentiating factor. While electric vehicles designed from the ground up are built on separate production lines and systems the conversion design approaches used by all German OEMs are produced on the same production lines as their conventional counterparts [1]. A particular challenge is the body assembly, because the production of a car body shell is currently carried out with rigid, interlinked systems, which are usually designed for one specific model (with derivatives of low variance in geometry and material) and designed for one operating point. The structure of a vehicle is highly dependent on the driveline used due to packaging, weight, safety, and stiffness requirements. Therefore, it is necessary to branch out the production line at production stages where different parts must be joined to different derivatives. This takes a considerable amount of floor space and increases the cost per vehicle as some of the branches will not be used at full capacity.

1.2   SmartBodySynergy

The aim of the BMBF-funded research project SmartBodySynergy is to develop a process for the integration of electric vehicles into existing production systems for conventional vehicles and to eliminate the described branching with low utilization in production. This is intended to ensure the economic production of car bodies for the model mix of conventional and electrified vehicles, which can be freely scaled to suit future developments. In order to achieve this, a paradigm shift in the car body assembly is necessary, analogous to the past transition from classic, rigidly interlinked transfer lines in machining to the use of agile production systems built up from parallel machining centers. The individual boxes are supplied from a parts supermarket using driverless transport systems as shown in Fig.1. The wbk Institute of Production Science in Karlsruhe is developing solutions for production control, quality assurance, and flexibility of the joining cells in cooperation with the project partners.

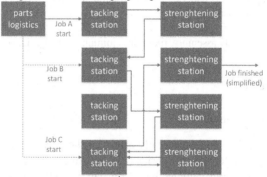

Fig. 1. Flow of a job through the body assembly in workshop production (one tacking station under maintenance)

With the aid of a new type of material flow and order control system, the degrees of freedom resulting from workshop production are used efficiently. This ensures a high utilization of the individual cells and a fast reaction to machine failures. Another challenge is the fact that the individual jobs can be carried out in variable order within the shell production facility, but must exit the process just-in-sequence. The order of the jobs when leaving the body assembly should be kept as precise as possible to keep the buffer tank small before surface treatment or final assembly. Possibilities for production planning and ways to react on disturbances are presented in [2]. In the area of quality assurance, concepts are drawn up at which points in the production quality assurance can be integrated and how the requirements placed on them change as a result of more flexible joining cells. Due to the more flexible material flow, it is possible, for example, to approach a quality assurance station with jobs from different joining stations in different degrees of completion. This allows a single station to be used to control the process at different points as required.

2 Fixtureless joining in car body assembly

2.1 Structure of the car body manufacturing

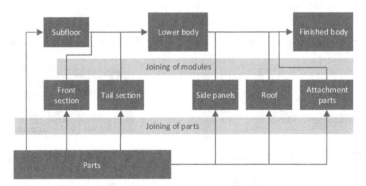

Fig. 2. Joining of parts and modules

Today, vehicle bodies are manufactured in linear, tree-like structures. First of all, the individual assemblies are combined into modules or module groups. These are then successively added to the car. The first modules are the front, underbody, and rear end of the vehicle, which are combined for assembly. The side walls, the roof and finally the additional parts are attached to these [3]. Fig. 2 shows the flow of a job through the traditional body assembly. The assembly of the modules will continue to be done conventionally within the framework of the project, but the individual modules will be produced in workshop production as described in Fig. 1.

| Positioning of the parts in a fixture | clamping | joining | Release of parts |

Fig. 3. Process steps in a single joining operation

In conventional joining stations, the joining partners are placed in rigid fixtures. The rigid fixtures determine the arrangement of the parts to each other and therefore must be very precise. The parts are then fixed with pneumatic clamping units and joined using robot-guided joining tools (Fig. 3). Due to the rigid fixtures, these stations are expensive and inflexible. Fixtureless joining is a possibility to assemble parts without the use of rigid and poorly-exchangeable floor mounted fixtures. Instead, the components are held in easily-exchangeable robot-guided grippers in all joining stations. On one hand, this is intended to reduce the costs for the installation of the joining cells. On the other hand, this allows the joining cells to become significantly more flexible as a result, which means that fewer cells and thus less space is required. The increased cycle time, due to the processing of several jobs on one cell and the conversion of the cell (switching of grippers and tools) between operations, might be acceptable due to the small lot size of the derivatives considered.

2.2 Fixtureless joining process

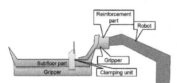

Fig. 4. Pre-positioning

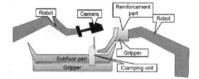

Fig. 5. Measuring the positioning error

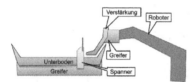

Fig. 6. Correcting the position

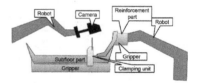

Fig. 7. Checking the position

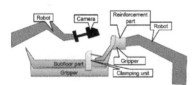

Fig. 8. Clamping the parts against each other

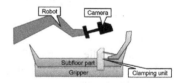

Fig. 9. Quality assurance

Industrial robots offer a sufficiently good repeatability (+/- 0.06mm) when positioning the grippers several times in the same pose within a short time [4]. Due to thermal expansion and other effects, however, the absolute positioning accuracy may be poorer when starting at different points in time [5, 6]. In order to compensate for deviations due to thermal expansion, robots can be temperature compensated [7, 8]. In robots executing measuring tasks, this is done by measuring ceramic measuring columns mounted in the cell and calculating the current model of the robot kinematics from the deviations of the measurement result and the known position of the ceramic columns. This makes it possible to use a robot-based camera to measure a wide variety of different positions in the cell with temperature compensation. In the concept of fixtureless joining, the actual position of the robots is measured with the aid of such a robot-guided camera system in order to compensate for deviations from the target position. For this purpose, the grippers with the parts are first brought to a pre-position in which there is still a safety distance between the parts. Then the exact position is measured and the remaining distance for the correct positioning of the parts is covered. If necessary, this process can be iterated. A similar procedure can be used for robot calibration [9]. After orienting, the position of the parts is checked again if necessary and, if the position is correct, the clamping units are activated. A quality assurance step can be installed before joining the parts. The alignment process is shown in Fig. 4 to Fig. 9. The feasibility of fixtureless joining for the production of spaceframe structures has been shown by [10]

3   Description of demonstrator task

Fig. 10. Underbody with reinforcement parts

The demonstrator cell shall clamp a reinforcement part to the rear carriage in the area of the rear wheel arch. The assembly is shown in Fig. 10. The process will be carried out as shown in Fig. 4 to Fig. 9. The basic principle of positioning and clamping requires a robot for part handling and a robot for measurements in accordance with the process of fixtureless joining. For space reasons, a robot is not used for the gripper of the larger part in the demonstrator. Instead, the gripper is placed on the floor without any precise positioning on low-stiffness legs. The correction of the gripper with the reinforcement part must therefore compensate for the positioning of the large part and the inaccuracies of the robot. At the end of the process, the parts are clamped together and the positioning accuracy is checked. No welding will be done due to safety reasons.

3.1   Robotic and measurement equipment in the manufacturing cell

The demonstrator cell is equipped with two Kuka KR180 R2500 robots. These robots were chosen because they have a load capacity and reach of 180kg and 2500mm, which is sufficient for handling car body parts. It should be noted that the manageable weight includes the grippers and depends on the distance between the center of gravity of the load and the robot flange, which is why some reserve in load capacity has been planned. The control system and the image analysis software run on a measurement computer. An SGS3D camera from ISRA VISION AG is used as the measuring system.

28

This is a camera with a built-in strip projector that can be used to determine the position of a component feature in the six dimensions. The camera system has been selected because it offers a good measuring accuracy of +/-0.1mm [11] and the measuring range is large enough to reliably place the component feature within it.

3.2 Gripper System

Fig. 11. Handling the parts in the grippers   **Fig. 12.** Parts positioned and clamped

Robot gripper systems are very widespread and are often used for handling tasks in automated production lines. They are usually composed of a frame, a mechanical coupling, and elements for repeatable positioning and clamping units. The grippers in the demonstrator unit are based on Industrial robot gripper systems. The grippers replace the rigid fixtures used in conventional body manufacturing. The gripper must therefore be able to take over the tasks of the fixture. These include the handling and positioning of the parts to be joined.

Another task that can be deduced from the project's goal is the possibility of clamping the reinforcements to the underbody as seen in Fig. 8. It can also be seen on the process description that the gripper holds the underbody on the underside. This is a good choice as there are several holes on the underside which are suitable for positioning. Furthermore, the underbody can be laid into the gripper from above and removed again after joining. This way of holding the component is also useful in terms of robot load. The short distance to the center of gravity reduces the moments acting on the robot hand, which means that the frame needs to be less rigid and is thus lighter in construction. Linearly sliding clamping units are used to enable the loading of the parts. The clamping units are moved away from the middle of the gripper to insert parts into the gripper. As soon as the parts are aligned with the camera, the clamping units move to the components and clamp them. The finished grippers are shown in Fig. 11 and Fig. 12.

3.3 Control architecture

To operate the flexible joining cell, all subsystems must communicate with each other. These are: the measuring system with SGS3D camera and the image evaluation software RoVis on a measurement computer, the two KRC4 controllers of the robots, and the grippers. To make this communication as flexible as possible, a "Communicator" written in the Python programming language is run on the computer, which establishes connections to the respective interfaces of the measuring system and the robots and controls the program sequence. The grippers in turn are operated directly via the robot controllers with a Profinet-capable bus coupler. The image analysis software RoVis has a TCP/IP-based interface which can be used for commissioning, feedback of operating conditions/errors, and for result transmission.

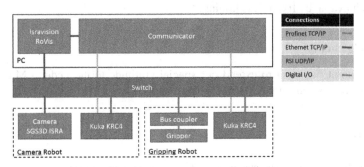

Fig. 13. Topology of the demonstrator station

The topology of the station is shown in Fig. 13. Part feature recognition is set up manually in the setup phase on the measurement computer. Drill holes on the components are selected as the component features used for positioning. After a pattern search in the entire measuring range of the camera using a best-fit comparison of a trained image of the feature, the center of the bore hole can be determined accurately in the second step of the search by means of a grey value-based edge search. With the help of line patterns projected by the SGS3D camera, triangulation can then be used to fit a plane onto the component, whereby the distance Z of the circle center point and the rotations Rx, Ry, Rz relative to the camera origin can be determined. These measured values are transmitted to the Communicator. With the positions of three points per component, the components position and orientation in space can be calculated with the help of the CAD model. A correction vector between the two components can then be calculated, which is transmitted to the gripping robot. The Robot Sensor Interface (RSI) is used for communication between the Communicator and the robot controllers. With an active RSI connection, data is sent back and forth between the communicator and robot controller in a configurable XML format using UDP/IP data in 12ms intervals. These variables can be used to transmit the previously calculated correction vector. Furthermore, by querying and setting these variables in the program of the robot controllers, it is possible to grant or revoke the go-ahead for travel, therefore allowing the Communicator to control the entire process flow. After the position of the parts has been corrected the robot control will close the clamping units mounted on the gripper. At the end of the process the robot handling the reinforcement part will release the part and move away. The clamped position of the parts can be controlled optically in a quality assurance step.

4   Summary

A concept for handling, positioning and clamping of two parts without the use of rigid fixtures has been presented in this paper. In this context a demonstrator cell has been developed and implemented. It shall be used for the flexible production of different derivatives of a vehicle and thereby enable an economic ramp up of electric vehicles. In the future the demonstrator station will be used to investigate the positioning quality achievable by fixtureless joining.

Acknowldgements

At this point, we wish to extend our special gratitude to the Federal Ministry for Economic Affairs and Energy (BMWi) for its support.

References

1. Brünglingshaus, C.: Fahrzeugkonzepte: Conversion versus Purpose Design. https://www.springerprofessional.de/fahrzeugtechnik/elektrofahrzeuge/fahrzeugkonzepte-conversion-versus-purpose-design/6561908 (2012). Accessed 12 November 2012
2. Echsler Minguillon, F., Lanza, G.: Maschinelles Lernen in der PPS. Konzept für den Einsatz von maschinellem Lernen für eine Matrix-Produktion. wt-online **107**(9), 630–634 (2017)
3. Pischinger, S., Seiffert, U. (eds.): Vieweg Handbuch Kraftfahrzeugtechnik, 8th edn. ATZ / MTZ-Fachbuch. Springer Vieweg, Wiesbaden (2016)
4. Kuka Roboter AG: Produktdatenblatt KR QUANTEC extra (2016)
5. Hesse, S.: Grundlagen der Handhabungstechnik, 4th edn. Hanser eLibrary. Hanser, München (2016)
6. Heisel, U., Richter, F., Wurst, K.-H.: Thermal Behaviour of Industrial Robots and Possibilities for Error Compensation, Universitat Stuttgart (1997)
7. Bongart, T.: Methode zur Kompensation betriebsabhängiger Einflüsse auf die Absolutgenauigkeit von Industrierobotern. Dissertation, Technische Universität München (2003)
8. ISRA VISION AG: In Line Messtechnik Alles aus einem Guss - optimal kombiniert. http://www.isravision.com/MESSTECHNIK-SCHL-uuml-SSELFERTIG------_site.site.html_dir._nav.2216_likecms.html. Accessed 11 October 2017
9. Barton, D., Schwab, J., Fleischer, J.: Automated calibration of a lightweight robot using machine vision (2017)
10. Elser, J.: Vorrichtungsfreie räumliche Anordnung von Fügepartnern auf Basis von Bauteilmarkierungen. Zugl.: Karlsruhe, Karlsruher Inst. für Technologie, Diss. Forschungsberichte aus dem wbk, Institut für Produktionstechnik, Karlsruher Institut für Technologie (KIT), vol. 180. Shaker, Aachen (2014)
11. ISRA VISION AG: Kompakte und universelle 3D Sensorik – SGS3D-BG +. Product Info. http://www.isravision.com/media/public/prospekte2014/sensorsystems/SGS3D_BG+_DE_2014-07_low.pdf (2014)

An Architecture for Intuitive Programming and Robust Execution of Industrial Robot Programs

Eric M. Orendt and Dominik Henrich

Chair for Applied Computer Science III
Robotics and Embedded Systems
University of Bayreuth, D-95440 Bayreuth, Germany,
E-Mail: {eric.orendt | dominik.henrich} @uni-bayreuth.de,
http://www.ai3.uni-bayreuth.de/members/orendt/

Abstract. Industrial robots are established tools in a variety of automation processes. An important goal in actual research is to transfer the advantages of these tools into environments in small and medium sized enterprises (SME). As tasks in SME environments change from time to time, it is necessary to enable non-expert workers to program an industrial robot in an intuitive way. Furthermore, robot programs must concern unexpected situations as many SME environments are unstructured, because humans and robots share workspace and workpieces. The main contribution of our work is a robot programming architecture, that concerns both aspects: Intuitive robot programming and robust task execution.
Our architecture enables users to create robot programs by guiding a robot kinesthetically through tasks. Relevant task information are extracted by an entity-actor based world model. From these information we encode the demonstrated task as a finite state machine (FSM). The FSM allows the reproduction and adaption of a task by the robot in similar situations. Furthermore, the FSM is utilized in a monitoring component, which identifies unexpected events during task execution.

Keywords: behavior based robotics, intuitive robot programming, robust task execution, entity actor framework, finite state machines, kinesthetic programming

1 Introduction and Related Work

While industrial robots are quite common in large assembly lines, we find only a small amount of them in domains with occasionally changing tasks, such as in small and medium sized enterprises (SMEs). SME domains enable new possible applications for robot systems and benefit from combining the advantages of robots and humans in a shared workspace [1]. From our point of view, there are two important issues which must be solved to establish industrial robots in SMEs. First, non-experts need to be enabled to create robot programs. Second, these programs has to be executed in a robust manner, regarding an unstructured and shared workspace.

The first point is related to *Programming by Demonstration (PbD)*, which is an intuitive programming paradigm, where a user demonstrates a task solution to the robotic system. Though there are numerous ways to create the required demonstration [2], our research focuses on kinesthetic PbD. This means that necessary program information

© Springer-Verlag GmbH Deutschland, ein Teil von Springer Nature 2018
T. Schüppstuhl et al. (Hrsg.), *Tagungsband des 3. Kongresses Montage Handhabung Industrieroboter*, https://doi.org/10.1007/978-3-662-56714-2_4

Fig. 1. (Left) Possible application area in an SME domain: A robot palletizes changing lots of loose products, while a human occasionally exchanges the product carrier. **(Right)** A user is intuitively programming a task to the robotic system by guiding it kinesthetically.

(like trajectory or gripper status) are extracted while a user guides the robot through a task. Most PbD approaches utilize methods from Machine Learning, which leads to the field of *Learning by Demonstration* [3]. They generate almost perfect reproductions from multiple demonstrations [4]. However, there is a niche for approaches which are based on single demonstrations (*One-Shot PbD* [5]), since some tasks have to be done quick (e.g. small lot sizes in SMEs) or it would be annoying work to give many similar performing examples [6]. As shown in [7], One-Shot PbD is rated as very intuitive for non-experts. There is just a small number of One-Shot PbD approaches, but they all match in the benefit of intuitive use [6, 5, 8–10]. The second point is related to the idea to detect unexpected events at a task based level. This motivation can be traced back to basic work on *model-based diagnosis* [11], which is an approved concept in the context of robotic task execution [12]. In general, these approaches require fault models to represent a relation between a failure and its possible causes. This means every single failure as well as the related causes must be defined, which is not suitable for occasionally changing tasks in SMEs. Additionally, the source of a failure is may not even detectable. Another basic approach is to use *monitoring* and compare an observed task solution to a prior known model of the observed task [13]. This allows to detect unexpected events by analyzing the differences between an observed and an expected sequence of a task model. In robotics research this has already been done for given task models [14, 15].

Our contribution is a architecture for kinesthetically programmable industrial robot systems, that enables non-experts to create, execute and monitor a robot program. We do this by combining components of previous work [15, 16, 10] and proposing an architecture, that utilizes an entity actor system to model a demonstrated task solution. This is done by observing and encoding workspace resources (e.g. objects or the gripper). Changes of these resources during the task demonstration are modeled in a finite state machine (FSM). By using FSMs we gain two advantages. First, it is possible to reproduce a task solution to a new, similar scene. Second, the same FSM can be used to detect unexpected situations during task execution. Furthermore it is possible to completely classify the occurring deviations as input for error recovery.

In the remainder of this paper, we present the architecture consisting of three components with its data structures and the used models. In the following we present details regarding the programming, execution and monitoring component. Subsequent we describe our experimental setup and evaluation results. Finally we draw conclusions and give an outlook to future work.

2 Architecture

This section gives a comprehensive overview of the proposed architecture, its components and the used data structures. In contrast, the following section elaborates on key details of the single components.

We go top-down by introducing first the two stages which determine the useable features of the architecture: Demonstration and execution stage. During *demonstration* a non-expert user gives a single example of a task solution in a certain robot workspace. This demonstration is captured by a sensor system including external (e.g. RGBD images) and internal parameters (e.g. joint angles). During *execution* it is the robots aim to solve a given task with the available information from the demonstration stage. It utilizes the same sensor system and also captures the programm execution.

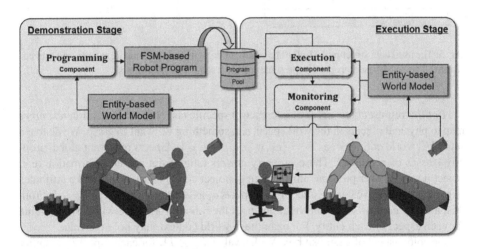

Fig. 2. Overview of stages, components, data structures and data flow in our suggested architecture. The programming component generates robot programs from given examples during demonstration stage. They can be applied and monitored during execution.

The architecture includes three active main components: The programming component, the execution component and the monitoring component. The *programming component* is part of the demonstration stage and generates executable robot programs from a given task example. The input for this component is a world model that represents a state sequence of the robots workspace during a demonstration. Its output is a finite state machine (FSM) that represents necessary robot actions and related workspace states.

Execution and monitoring component are part of the execution stage. The *execution component* selects a program from a pool of possible FSM programs created by the programming component. The selection is based on the actual world models state and the output are executable robot controller instructions. The *monitoring component* observes the execution of these instructions by comparing the expected workspace changes (retrieved from the execution component) and the actual captured workspace changes (retrieved from the world model).

The underlying data structures are a world model based on the ENACT framework [16] and a finite state machine. The *world model* represents the robots workspace by so called world contexts and consists of entities, aspects and actors as shown in Fig. 3.

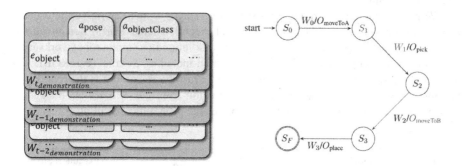

Fig. 3. The used data structures: **(Left)** The world model with world contexts, entities and aspects. **(Right)** The FSM-based program representation with states and operations.

Entities represent resources of interest to a specific task. In our application *resources* map to physical objects in the workspace, e.g. something we want to grasp. We denote a set of $|E|$ world entities e_i as $E = \{e_1, e_2, e_3, \ldots, e_{|E|}\}$. *Aspects* represent global property classes of resources. These property classes range from spatial information (e.g. object position) over physical attributes (e.g. object color) to knowledge-like attributes (e.g. object type). We denote a set of $|A|$ aspects a_i as $A = \{a_1, a_2, a_3, \ldots, a_{|A|}\}$. From this, a *world context* stores information about the robots workspace with a single value for each aspect of each entity. For example, a world context may hold the value "'red'" for the color aspect of a deposit box. With a value $d_{i,j} \in D_j$ for each aspect a_j in entity e_i we can denote a world context W as

$$W : E \to D_1 \times D_2 \times D_3 \times \cdots \times D_{|A|},$$
$$W(e_i) = (d_{i,1}, d_{i,2}, d_{i,3}, \ldots, d_{i,|A|}).$$

Actors represent active elements, which update a world context over time. For instance, position aspects of objects may change during a task demonstration and a sensor actor updates these. Formally each actor $\mathfrak{a}_k$ from a set of $|\mathfrak{A}|$ actors $\mathfrak{A} = \{\mathfrak{a}_1, \mathfrak{a}_2, \mathfrak{a}_3, \ldots, \mathfrak{a}_{|\mathfrak{A}|}\}$ works on entities $E_{\mathfrak{a}_k} \subset E$ and their aspects.

The *FSM* represents a robot program, which is usually a task solution or a part of it. Formally we denote an FSM as a 6-tuple $B = [S, W, O, \zeta, S_0, S_F]$ with a set of states

S, an initial state $S_0 \in S$, a final state S_F, a set of world-contexts $W = \{W_0, \ldots, W_n\}$ with $W_i = \{W_i(e_0), \ldots, W_i(e_{|E|})\}$ and a set of robot operations $O = \{O_0, \ldots, O_n\}$ with $n \in \mathbb{N}^+$. The state-transition function ζ is defined by $\zeta : S \times W \rightarrow S \times O$. A program B follows a sequence of operations: To realize a transition from state S_a to state S_b a certain world context W_a must be present. This combination triggers a robot operation O_a (e.g. a P-to-P movement or a grasp), which results in a new world context. When this new world context corresponds to W_b, we bring on the next operation O_b. For instance, Fig. 3 illustrates a typical FSM representation for a pick-and-place task.

Further details as well as advantages and disadvantages of the chosen data structures can be found in [15, 16] and [10].

3 Components

This section provides details on the proposed components. Due to limited scope we refer to original sources where appropriate.

3.1 Programming component

The main goal of the programming component is to extract a robot program from a given example during the demonstration stage. This component is based on the insights and results of [10]. To generate a FSM we extract components of B from a demonstrated task. Most of the components in B are directly extracted from ENACT: W is a subset $W \subseteq \mathcal{W}$, while there is a state for every element in W. The initial and final state S_O and S_F correspond to the first and last world context W_{t_0} and $W_{t_{end}}$.

It is more complex to extract the robot operations O. This corresponds to the question which subset of $\mathcal{W}$ exactly is chosen for W. To solve this, we first introduce static and dynamic aspects. A *dynamic aspect* is often part of the robotic system and can be changed directly by an operation $O_i \in O$. For instance, the position of picked objects or the gripper status (open, closed) are dynamic aspects. *Static aspects* instead are not changeable. For instance, the color of a workpiece (except in a painting scenario). The classification whether an aspect is dynamic or not can be done by checking if an aspect a_j in entity e_i has more than one value $d_{i,j} \in D_j$ in all occurred world contexts $\mathcal{W}$.

From this point, it is easy to see that robot operations are actors in the ENACT framework as they change aspects as well. Though we can form the specific robot operations $O = \{O_0, \ldots, O_{|O|}\}$ by analyzing the dynamic aspect changes between two consecutive world contexts W_t and W_{t+1} and define actors

$$\mathfrak{a}_t, c : W_t(e_i) = U_{\mathfrak{a}_{t,c}}(W_{t-1}, e_i) \forall i \in (1, \ldots, |E|)$$

that carry out the observed aspect changes and thus are mapped to robot specific operations $O_t : W_t \rightarrow W_{t+1}$, which determine states $S = \{S_0, \ldots, S_t, S_{t+1}, \ldots, S_{t_{end}}\}$. In our prototype setup we use 3 elementary operations: A pick (down-and-upward path with closing gripper), a place (down-and-upward path with opening gripper) and a move operation (point to point path).

3.2 Execution Component

The goal of the execution component is to generate executable code. From the programming component we obtain an FSM (representing a robot program), which produces valid robot operations as a transition output between states as introduced in [10].

As a major advantage this concept directly grants a sequence of low-level robot operations. Every transition of the FSM represents a joint angle or TCP movement, or an end effector operation (e.g. grasp). Hence the state machine is without branches or loops, the operations $\{O_0, \ldots, O_{t_{end}}\}$ are rolled out directly as illustrated in Fig. 4.

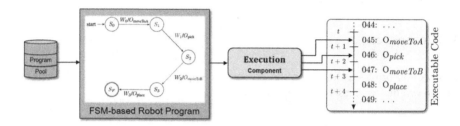

Fig. 4. When a matching robot program is found, the execution component maps FSM-based representation to executable robot code.

The core idea is, that the robotic system is able to switch between programs to solve different tasks. The programs itself are executed in a linear way. The decision which program is selected is based on the actual world model. This is done by comparing the actual world state with the initial states S_0 of each FSM in the program pool. When there is a matching program, the programming component can start to execute it. Note, that there should be a strategy to handle situations with more than one matching program.

3.3 Monitoring Component

The goal of the monitoring component is to observe the execution of a robot program and identify unexpected events. This is an integrated module from previous work [15]. It is based upon entity-based resources and hence easily to apply during the task execution. Since it is a comprehensive approach we only present core details at this point.

Our key idea is to take advantage from world contexts $\mathcal{W} = \{W_{t_0}, \ldots, W_{t_{end}}\}$ we obtain during task demonstration. In the same manner we obtain a second set of world contexts $\mathcal{W}^R = \{W^R_{t_0}, \ldots, W^R_{t_{end}}\}$ during task reproduction. Following the definitions from Sections 2 and 3.1 we have on one hand the *expected* world-context W_t and the *observed* world context W^R_t on the other hand. According to the representing FSM we can distinguish two cases.

Either expected and observed world context are equal, i.e. $W_t = W^R_t$. In this case the operation was successful. The FSM can perform a transition from S_t to S_{t+1} and task execution goes on. Or expected and observed world context are unequal, i.e. $W_t \neq W^R_t$. In this case the operation was unsuccessful. The FSM remains in state S_t. From this we can detect deviations during execution of FSM-based programs. According to [15] we identify 5 different deviation types by analyzing affected entities regarding their change from one state to another:

1. The existence of an entity $e_i \in E$ is expected, but no matching e_i^R is observed.
2. The absence of an entity $e_i \in E$ is expected, but a matching e_i^R is observed.
3. The value $d_{i,j}$ of an aspect a_j in entity e_i is expected to be changed after a time step with $W_{t+1}(e_i) = U_{a_{t+1,c}} W_t(e_i)$, but no change occurs $(W_{t+1}^R(e_i) = W_t(e_i))$.
4. The value $d_{i,j}$ of an aspect a_j in entity e_i is expected not to be changed after a time step with $W_{t+1}(e_i) = W_t(e_i)$, but a change occurs $(W_{t+1}^R(e_i) = U_{a_{t+1,c}} W_t(e_i))$.
5. The value $d_{i,j}$ of an aspect a_j in entity e_i is expected to be changed with $W_{t+1}(e_i) = U_{a_{t+1,c}} W_t(e_i)$, but another change occurs $(W_t(e_i) \neq W_{t+1}^R(e_i) \neq U_{a_{t+1,c}} W_t(e_i))$.

The deviation types are complete and unified, that means any unexpected situation can be unambiguously assigned to one type. This grants a useful basis for error handling, because every type demands specific recovery strategies. For instance, deviations from Type 2 can often be solved by simply repeat an operation, while deviations from Type 5 need a more complex strategy. Note, that the proposed component can only detect and identify unexpected situations, when they are task related and thus affect entities.

4 Experiments

We evaluated our components with an experimental setup as shown in Fig. 1. It consists of a parallel gripper PG70 from SCHUNK, which is mounted to a KUKA Lightweight Robot. A Kinect depth sensor above the workbench delivers frequently point clouds to observe the workspace (i.e. entities and aspects) by using the Point Cloud Library [17].

Our experiments were embedded in a larger user study [7] and focused on two aspects: Investigating on the intuitiveness of the programming approach and validating the FSM-based program execution as well as the monitoring. In particular, the participants are experts from the field of robotics and non-experts without relevant prior knowledge.

For evaluation we performed benchmark tasks from [7]. Task 1 is to move a building block from a freely chosen start position to a deposit area. Task 2 is to disassemble a tower of two building blocks and rebuild it upside down. Task 3 is to sort building blocks into a trash box or to a deposit area, depending on their color.

In a first experiment [10], the first and second task were demonstrated 50 times by different users with freely chosen positions in the workspace. Each demonstration was executed 3 times, resulting in 300 executions. 89% of them were successful. Failed executions seem to be caused by inaccurate demonstrations: 27 of 32 failed reproductions belong to a demonstration, in which all executions failed. A less common reason for fails are software (e.g. false perception) and hardware issues (e.g. striking gripper), which are out of scope and task unrelated. Without these issues, all occurring deviations were detected correctly with a noticeable distribution of deviation types: Most of them (43% and 29%) were from Type 5 and 4. This was usually caused by an inaccurate place operation, e.g. when a building block was released too high or pushed to hard against the other block during stacking.

38

In a second experiment [7] we evaluated the intuitiveness of our kinesthetic one-shot approach with the Questionnaire for the subjective consequences of inuitive use (QUESI) according to [18]. Overall 34 participants were asked, which created 102 programs. The results are illustrated in Fig. 5.

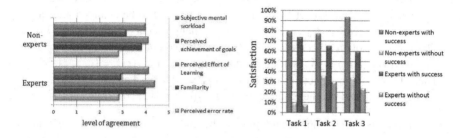

Fig. 5. Resulting user satisfaction according to QUESI (**left**) and related to the programmers success (**right**). Higher values represent better results. [7]

We evaluated a QUESI score of 3.6 for experts as well as for non-experts. A study, which analyzed the intuitive usability of different interaction methods for a smart device game, evaluated a QUESI score of 4.0 for the use of touchscreen [19]. The comparison of these values results in a difference of 0.4. Considering that most of the participants in [19] use smart devices regularly and our non-experts never used a robot before, this result is quite promising.

Summarizing, the experiments validate the principle and usability of the system and confirms the intuitive of use. However, additional measures are necessary to examine the architectures capabilities in different applications and SME domains.

5 Conclusions & Future Work

This paper presents an architecture that enables non-experts to use the capabilities of a general purpose industrial robot. The programming component allows to generate robot programs modeled by a FSM without prior expert knowledge. The execution component applies the generated programs to similar situations in the workspace to solve a shown task. During this execution, the monitoring component detects and identifies unexpected deviations regarding the task fulfillment. Consequently, our future work may deal with a recovery component to handle identified deviations.

We validated our approach in experiments, which shows promising results. However, our architecture is not limited to the shown tasks and can be applied to different tasks in non-expert domains by customizing entities and aspects of the underlying world model. Summarizing our contribution carries significance for applications from the field of intuitive robot programming, robust task execution and human-robot collaboration.

Acknowledgements. This work has been supported by the Deutsche Forschungsgemeinschaft (DFG) under grant agreement He2696/15 INTROP.

References

1. T. Werner, *et al.*, "Design and Evaluation of a Multi-Agent Software Architecture for Risk-Minimized Path Planning in Human-Robot Workcells," in *Tagungsband des 2. Kongresses Montage Handhabung Industrieroboter*, vol. 5, 2017, pp. 1–10.
2. B. D. Argall, *et al.*, "A survey of robot learning from demonstration," *Robotics and Autonomous Systems*, vol. 57, no. 5, pp. 469–483, 2009.
3. C. Atkeson and S. Schaal, "Robot learning from demonstration," *Machine Learning*, no. 1994, pp. 12–20, 1997.
4. A. Billard and S. Calinon, "Robot Programming by Demonstration," in *Handbook of Robotics, Chapter 59*, 2007, vol. 48, pp. 1371–1394.
5. Y. Wu and Y. Demiris, "Towards one shot learning by imitation for humanoid robots," in *Proceedings of IEEE ICRA*, Anchorage, 2010, pp. 2889–2894.
6. R. Zöllner, *et al.*, "Towards cognitive robots: Building hierarchical task representations of manipulations from human demonstration," in *Proceedings of IEEE ICRA*, Barcelona, 2005, pp. 1535–1540.
7. E. M. Orendt, *et al.*, "Robot Programming by Non-Experts: Intuitiveness and Robustness of One-Shot Robot Programming," in *Proceedings of IEEE RO-MAN*, New York, 2016.
8. B. Akgun, *et al.*, "Keyframe-based Learning from Demonstration: Method and Evaluation," in *Proceedings of Social Robotics*, Chengdu, 2012, pp. 343–355.
9. C. Groth and D. Henrich, "One-Shot Robot Programming by Demonstration using an Online Oriented Particles Simulation," in *Proceedings of IEEE ROBIO*, Bali, 2014.
10. E. M. Orendt, *et al.*, "Robust One-Shot Robot Programming by Demonstration using Entity-based Resources," in *Advances in Service and Industrial Robotics*, Turin, 2017.
11. R. Reiter, "A theory of diagnosis from first principles," *Artificial Intelligence*, vol. 32, no. 1987, 1987.
12. G. Steinbauer and F. Wotawa, "Robust Plan Execution Using Model-Based Reasoning," *Advanced Robotics*, pp. 1315–1326, 2009.
13. V. Verma, *et al.*, "Real-time fault diagnosis," *IEEE Robotics and Automation Magazine*, vol. 11, no. June, pp. 56–66, 2004.
14. A. Kuestenmacher and E. Al, "Towards Robust Task Execution for Domestic Service Robots," *Journal of Intelligent & Robotic Systems*, pp. 528–531, 2014.
15. E. M. Orendt and D. Henrich, "Design of Robust Robot Programs: Deviation Detection and Classification using Entity-based Resources," in *Proceedings of IEEE ROBIO*, Zhuhai, 2015.
16. T. Werner, *et al.*, "ENACT: An Efficient and Extensible Entity-Actor Framework for Modular Robotics Software Components," in *International Symposium on Robotics*, Munich, 2016, pp. 157–163.
17. R. B. Rusu and S. Cousins, "3D is here: point cloud library," in *Proceedings of IEEE ICRA*, Shanghai, 2011, pp. 1 – 4.
18. A. Wegerich, *et al.*, *Handbuch zur IBIS Toolbox*. Technische Universität Berlin, 2012.
19. F. Pollmann, *et al.*, "Evaluation of interaction methods for a real-time augmented reality game," *Lecture Notes in Computer Science*, vol. 8215 LNCS, pp. 120–125, 2013.

Augmented Reality for teaching collaborative robots based on a physical simulation

Patrick Rückert[1], Finn Meiners[1], Kirsten Tracht[1][0000-0001-9740-3962]

[1] University of Bremen, Bremen Institute for Mechanical Engineering, Badgasteiner Straße 1,
28359 Bremen, Germany
rueckert@bime.de

Abstract. Augmented reality as a technology for programming a collaborative robot is in comparison to existing practice a more intuitive, safe and quick approach. This paper shows a novel concept for connecting a physical robot with its virtual twin, using augmented reality as a visual output and a device for interacting with the simulation model. The core of the system is a station control that contains all interfaces, sequencing logics and databases. For programming the robot, the coordinates are created by the operator by interaction and voice input in the augmented reality environment. All information is stored in a shared database which contains assembly plans and instructions for both workers and robots. A demonstration platform is presented, that allows to implement the concept of the system design for an experimental assembly station for human-robot collaboration. The work station includes a Universal Robot 5® that assists the worker in handling of objects and is operated by coordinates from the station control. A Microsoft Hololens® is used for programming the robot in an augmented reality environment which is based on the output of the physical robot and the work system simulation.

Keywords: Augmented Reality, Simulation, Robotics.

1 Introduction

The increase of mass customization requires high flexibility in assembly systems for both humans and robots. Collaborative robots, that are designed for assembly lines, can support humans by reducing physical and cognitive load, while improving safety, quality and productivity. The main enablers for this technology are intuitive interfaces, safe strategies and equipment, proper methods for planning and execution, the use of mobile robots and distributed computing [1, 2]. A key technology for implementing these factors are elements of mixed reality that can improve manufacturing processes by giving visual information and enabling interaction with manufacturing systems [3]. Augmented Reality (AR) as a technology of mixed reality has a high potential for robot programming, due to the high interaction between robot, model and object [4]. This paper introduces a concept for a system design for using AR as a robot programming tool while accomplishing the mentioned benefits.

© Springer-Verlag GmbH Deutschland, ein Teil von Springer Nature 2018
T. Schüppstuhl et al. (Hrsg.), *Tagungsband des 3. Kongresses Montage Handhabung Industrieroboter*, https://doi.org/10.1007/978-3-662-56714-2_5

2   Augmented reality for programming industrial Robots

As robot programming systems move beyond basic, text-based languages and record-and-play programming, they are becoming significantly more powerful. More intuitive methods that reduce the work which is required to implement systems allow people with little or no technical skills to program robots [5]. The increasing use of various robotic systems and programming methods raises the need to enhance their compatibility. A standardized technology for programming and controlling different robot system is not implemented for the industry yet [6].

One approach for supporting especially unskilled workers with programming robots for collaborative assembly tasks is the programming by demonstration. With support of a graphical programming interface, the user can record the demonstrated tasks and the specific programming code is created automatically and adapted for the specific robot system [7]. The intuitiveness of the programming method is a factor that enables workers to teach a robot after only a short instruction. This approach can be further expanded by the use of gesture recognition instead of moving the robot by hand or control panel. The hand position and movement can be recorded by a depth camera and transferred into coordinates. A client-server based station controller handles the coordinates and executes the movement of the robot [8].

A graphical interface for programming robots online can be supported by information from a simulation-programming tool, which generates the program structure of the robot. The robot program code is uploaded through a graphical interface and the program structure is generated automatically. The user can interact online with a program via gestures and voice commands even if it was generated by the offline tool [9, 10].

A method for combining the intuitiveness for quick and simple applications and programming robots safely is the use of AR [11]. The operator is equipped with proper interaction tools, for example a head-up display and interacts intuitively with the provided information during the robot programming process. Using virtual robot models or simulations enables unskilled workers to program a robot by guiding the virtual models without having to interact with the real robot at the assembly station [9]. A method for 2D path planning is a laser-projection-based approach where the planned paths are projected over the real work piece and the operator can manually edit and modify them through an interactive stylus [12,13]. For 3D AR path planning, collision-free paths of a robot can be generated automatically by using a beam search algorithm based on free spaces and start and goal definitions set by the worker [11]. A concept that allows the user to execute different tools for programming, is the implementation of a multimodal robot programming toolbox. The system contains both online and offline methods and is connected to an AR device [14].

By using 3D simulations for design and development of production systems, the software quality can be increased and the ramp-up time can be shortened [15]. Using AR devices, modification costs after the installation process can be decreased by avoiding adjustment and optimization, resulting in a reliability of 98% if the AR-analysis is performed in collaboration with the construction engineer. [16]

The system design which is presented in this paper picks up the presented concepts of programming by demonstration, voice and gesture control, offline simulation and AR and transfers them to an intuitive method for programming collaborative robots in AR environment. The developed system differentiates itself from the existing solutions by combining a simulation software for production processes in mechatronic production with concepts of worker assistance, allowing both, the robot operator and the production engineer, to use a common system.

3 System design and concept

The idea behind the system design is the connection of physical robot manipulator with the simulation of its virtual twin by using a station controller that is operating the robot by sending and receiving coordinates. The virtual twin is a model of the robot which represents the kinetic chain, physical body and all degrees of freedom. Additionally, it is possible to add a gripper to the simulation or extend it with other kinematics from the workspace that are needed to resemble the process.

The AR headset is implemented as a real-time visual output of the simulation and enables the operator to send commands and interact with the virtual robot. For merging the simulation with the reality, the user can set up the virtual robot at the position of the physical robot as shown [Fig. 1]. The model of the robot is adapted from the virtual twin of the workstation which is realized with the Simulation Software industrialPhysics®. This enables two programming methods. The first one uses the simulation of the workstation [Fig. 3] for programming the robot virtually and transferring the program to the physical workstation. Another option and focus of the presented work is the programming by AR by using either voice control or gestures to interact with the robot. The benefit is, that the operator can interact directly with a model of the robot at the work space and detect possible collisions, safety risks or non-ergonomic assembly positions.

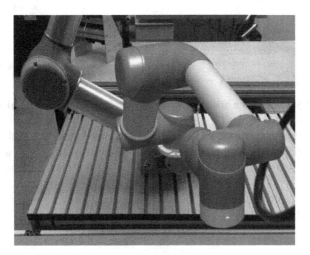

Fig. 1. Merging of augmented robot with the position of the physical robot

The main execution system, that is coordinating, monitoring and executing all interfaces, databases and the sequences of human and robot tasks, is called station controller [Fig. 2]. This includes the graphical user interface for the worker and administrator, that allows to start a collaborative task or to create new assembly plans. An interface to the database is used to store and receive all information from a SQL database which includes all assembly plans, robot coordinates and individual worker data. The assembly plans from the database include both worker and robot instructions and need to be complemented with robot coordinates. The coordinates can be either implemented by classical teach in or by using the implemented method of AR programming. The coordinates that are created from the user during AR programming are sent to the simulation and stored in the database by the hardware controller, which is used for the connection between the robot and the control station. The output of the instructions from the assembly plans is handled by a sequence logic that passes the instructions to worker or robot.

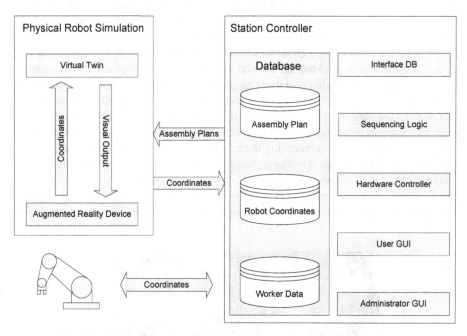

Fig. 2 System design of station controller and physical robot simulation

After the programming process, all needed data are stored in the assembly plan and can be retrieved during the assembly process either by the worker through a visual output on a touch display or by the robot through coordinates or gripper actions, that are send by TCP/IP connection. The AR headset is only needed for teaching the robot and a connection to the physical simulation is not necessary after the programming of the robot is finished.

4   Concept for application on a demonstration platform

The presented design of the system can be applied to an experimental assembly station for human-robot collaboration [Fig. 3]. The existing work station includes a Universal Robot 5® that assists the worker with the handling of objects and is operated by coordinates from the station control. A Microsoft Kinect® is used to implement different assistance tools for the worker, such as an adaption on height and reach and a progress monitoring of the assembly process [17]. All interaction with the system can be done either by voice, gestures or touch display. The control system of the worker assistance tool is fully integrated with an existing database structure for robot and worker instructions. For integrating the developed system design, the station control can be connected with a physical simulation, based on the software industrialPhysics®. The connection can be established by giving the simulation access on the database and by using and integrating a technology plugin based on C++. As an AR headset, the Microsoft Hololens® can be used, since it includes the native integration of voice and gesture recognition.

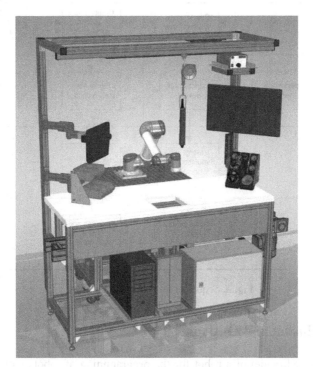

Fig. 3. Demonstration platform

A programming sequence can be implemented for the demonstration platform which is divided into operation paths for the station control, physical simulation, AR device and worker [Fig. 4]. As a first step for programming the robot, the operator logs himself in at the station control and creates a new assembly plan or opens an existing one. After creating a new robot instruction, the system is put on hold and waits for coordinates from the simulation that are written into the database. In parallel, the physical robot simulation is started and the virtual twin of the robot is set up. The simulation is connecting with the Microsoft Hololens®, sending the output information and waiting for input data. As a next step the virtual robot is calibrated to fit into the environment of the workspace. As soon as the station control and the simulation are ready for the programming of a new robot position, the user gets a signal on the AR headset and can start the programming of the robot. The programming itself is operated by intuitive voice and gesture commands. After the virtual robot reached the programmed position, the coordinates are transferred to the simulation and stored in the database of the station control as coordinates of robot instructions of the active assembly plan. After that, the loop starts again and a new robot instruction can be programmed until all robot instructions are set up.

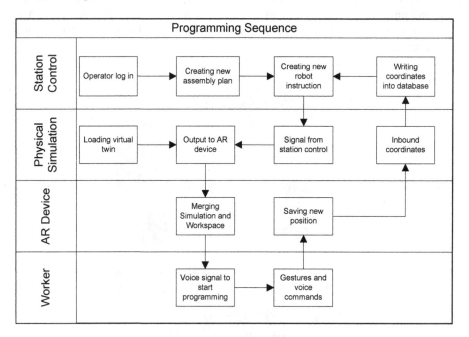

Fig. 4. Sequence for robot programming

5   Conclusion and Outlook

Using a simulated model of a robot for the programming of robots with AR devices opens up a variety of new opportunities for programmer, production planner and operator.

A huge benefit of the separation between physical robot and virtual twin is the programming of the robot without interfering with the actual production process. The programmer can implement, optimize and test new program sequences without having any interaction with the physical model of the robot that is running in production.

The application of a physical simulation allows to extend the simulation by other factors that could influence the path and motion of the robot, such as different grippers, transport times or cycle time of surrounding processes. By integrating these objects into the augmented environment, influences can be examined without the risk of physical crashes. A concept that involves human, physical objects and virtually simulated objects is one of the biggest potentials for the application of augmented robot programming, that could bring a huge benefit for production planers.

From the viewpoint of the operator, a tool is offered which supports him in programming a robot without actual proficiency in creating program code. An intuitive operation of the system makes it possible for an unskilled worker to interact with the robot. The implementation of coordinates into assembly plans, gives the advantage of having just one system running at the workstation, that the worker can easily handle without changing between different interfaces that differ in their complexity. Programming a virtual model of the robot should also lower the inhibition threshold of the worker regarding the chance of physically damaging the system by making mistakes in programming and handling the robot. This also allows to qualify and teach workers, who never have been interacting with a robot before and lower the fear of collaborative assembly. A topic that must be addressed in future research, is the gap between simulation and real assembly system. The programming of the robot can only be as exact as the calibration between simulation and physical work space. Therefore, pick and place application can only be realized by adding additional systems for calibration and machine vision, such as depth cameras. The deviations between virtually planned program sequences and the actual robot movement must be addressed after the fully implementation of the system to evaluate the precision of the calibration and the voice and gesture control. Another upcoming research topic is the evaluation of the programming system with production workers. Acceptance, intuitiveness and ease-of-use of the system still needs to be evaluated and compared to other state of the art programming methods.

References

1. Andrea Cherubini, A.; Passama, R.; Crosnier, A.; Lasnier, A.; Fraisse, P.: Collaborative manufacturing with physical human–robot interaction. In: Robotics and Computer-Integrated Manufacturing, Volume 40, pp. 1-13, (2016).
2. Michalos, G.; Makris, S.; Papakostas, N.; Mourtzis, D.; Chryssolouris, G.: Automotive assembly technologies review: challenges and outlook for a flexible and adaptive approach. In: CIRP Journal of Manufacturing Science and Technology, 2 (2), pp. 81-91, (2010).
3. Novak-Marcincin, J.; Barna, J.; Janak, M.; Novakova-Marcincinova, L.: Augmented Reality Aided Manufacturing. In: Procedia Computer Science, Volume 25, pp. 23-31, (2013).
4. Eschen, H.; Kötter, T.; Rodeck, R.; Harnisch, M.; Schüppstuhl, T.: Augmented and Virtual Reality for Inspection and Maintenance Processes in the Aviation Industry. In: 6th International Conference on Through-life Engineering Services, TESConf, (2017).

5. Biggs, G., MacDonald, B.: A Survey of Robot Programming Systems. In: Australasian Conference on Robotics and Automation, pp. 1-10, (2003).
6. Magaña, A.; Reinhart, G..: Herstellerneutrale Programmierung von Robotern - Konzept einer Softwarearchitektur zur Steuerung und Programmierung heterogener Robotersysteme. In: wt Werkstattstechnik online, 9, pp. 594-599, (2017).
7. Brecher, C.; Göbel, M.; Herfs, W.; Pohlmann, G.: Roboterprogrammierung durch Demonstration - Ein ganzheitliches Verfahren zur intuitiven Erzeugung und Optimierung von Roboterprogrammen. In: wt Werkstattstechnik online. 9, pp. 655-660, (2009).
8. Chen, S., Ma, H., Yang, C., & Fu, M.Hand gesture based robot station controller using leap motion. In: International Conference on Intelligent Robotics and Applications, Springer International Publishing, pp. 581-591, (2015).
9. Makris, S.; Tsarouchi, P.; Matthaiakis, A. S.; Athanasatos, A.; Chatzigeorgiou, X.; Stefos, M.; Aivaliotis, S.: Dual arm robot in cooperation with humans for flexible assembly. CIRP Annals-Manufacturing Technology, 66, pp.13-16, (2017).
10. P. Tsarouchi, S. Makris, G. Michalos, A.S. Matthaiakis, X. Chatzigeorgiou, A. Athanasatos, M. Stefos, P. Aivaliotis, G. Chryssolouris, G.: ROS Based Coordination of Human Robot Cooperative Assembly Tasks—An Industrial Case Study. In: Procedia CIRP, 37, pp. 254-259, (2015).
11. Chong, J.W.S., Ong, S.K., Nee, A.Y.C., Youcef-Youmi, K.: Robot programming using augmented reality: an interactive method for planning collision-free paths. In: Robot. Comput. Integr. Manufactoring 25(3), pp. 689–701, (2009).
12. Neea, A.Y.C., Onga, S.K., Chryssolourisb, G., Mourtzisb, D.: Augmented reality applications in design and manufacturing. In: CIRP Annals - Manufacturing Technology 61, pp. 657-679, (2012).
13. Zäh, M. F.; Vogl, W.: Interactive Laser-projection for Programming Industrial Robots. In: Proceedings of the International Symposium on Mixed and Augmented Reality, Santa Barbara, CA: 22–25 October 2006, pp. 125-128, (2006).
14. Andersson, N.; Argyrou, A.; Nägele, F.; Ubis, F.; Esnaola Campos, U.; Ortiz de Zarate, M.; Wilterdink, R.: AR-Enhanced Human-Robot-Interaction - Methodologies, Algorithms, Tools. In: Procedia CIRP, Volume 44, pp. 193-198, (2015).
15. Wünsch, G.: Methoden für die virtuelle Inbetriebnahme automatisierter Produktionssysteme. Zugl.: München, Techn. Univ., Diss., 2007. München: Utz Forschungsberichte / IWB, 215, (2008).
16. Pentenrieder, K.; Bade, C.; Doil, F.; Meier, P.: Augmented reality-based factory planning - an application tailored to industrial needs. In: ISMAR'07: Proc. 6th Int'l Symp. on Mixed and Augmented Reality, pp. 1–9, (2007).
17. Rückert, P.; Adam, J.; Papenberg, B.; Paulus, H.; Tracht, K.: Assistenzsysteme zur kollaborativen Montage - Mensch-Roboter-Kollaboration basierend auf Bildverarbeitung und Sprachsteuerung. In: wt Werkstattstechnik online, 9, pp. 556-571, (2017).

Visualizing trajectories for industrial robots from sampling-based path planning on mobile devices

Jan Guhl[1], Axel Vick[2], Vojtech Vonasek[3] and Jorg Krüger[2]

[1] Department of Industrial Automation Technology, Technische Universitat Berlin, Germany
[2] Fraunhofer Institute for Production Systems and Design Technology (IPK) Berlin, Germany
[3] Faculty of electrical engineering , Czech Technical University Prague, Czech Republic
guhl@iat.tu-berlin.de

Abstract. Production lines are nowadays transforming into flexible modular and interconnected cells to react to rapidly changing product demands. The arrangement of the workspace inside the modular cells will vary according to the actual product being developed. Tasks like motion planning will not be possible to precompute. Instead, it has to be solved on demand. Planning the trajectories for the industrial robots with respect to changing obstacles and other varying environment parameters is hard to solve with classical path planning approaches. A possible solution is to employ sampling-based planning techniques. In this paper we present a distributed sampling-based path planner and an augmented reality visualization approach for verification of trajectories. Combining the technologies ensures a confirmed continuation of the production process under new conditions. Using parallel and distributed path planning speeds up the planning phase significantly and comparing different mobile devices for augmented reality representation of planned trajectories reveals a clear advantage for hands-free HoloLens. The results are demonstrated in several experiments in laboratory scale.

Keywords: Industrial Robots, Path Planning, Augmented Reality

1   Introduction

The factories of the future will be characterized by drastically changing products and production processes within a very short time. That will lead to shorter product life cycles and demand for highly flexible and efficient production systems. One of these systems is an industrial robot that is capable of automatic adaption to its environment rather than being reprogrammed with great effort. The changing environment means, in first place, the appearance and vanishing of obstacles of any kind. Assuming the possibility to detect these obstacles by advanced methods of computer vision, it is, however, not trivial to react to this knowledge automatically. Sampling-based path planning methods compute new trajectories by sampling configurations around a former path randomly and checking for collisions regularly. Unfortunately, these methods demand for an immense computing power, what makes them unsuitable for fast re-planning. Using parallel and distributed path planning in the cloud overcomes

© Springer-Verlag GmbH Deutschland, ein Teil von Springer Nature 2018
T. Schüppstuhl et al. (Hrsg.), *Tagungsband des 3. Kongresses Montage Handhabung Industrieroboter*, https://doi.org/10.1007/978-3-662-56714-2_6

this restriction. The second disadvantage oJ the sampling-based planningJ the lack oJ determinism due to randomnessJ will be compensated by allowing the veriJication oJ a new trajectory by the human operator through the augmented reality visualization techniques.

2   Related Work

2.1   Distributed motion control for industrial robots

The classical industrial robot consists of a hardware platform with six or seven degrees of freedom and a control system in a machine cabinet next to it. Within the evolution of production systems and the change from hierarchical to interconnected factory control, the robot control system is no longer strictly fixed to its hardware. Centralized or distributed IT infrastructures like Clouds are used to outsource control computation and to connect machines and robots.

The technological challenges in computation, communication and security when outsourcing robotics function to the cloud are discussed in [Hu2012]. The authors use an ad-hoc cloud between several robots to evaluate communication protocols and show the capabilities of networked robotics. The feasibility of integrating value-added services to the field of cloud robotics is shown by [Dyu2015]. The Infrastructure- and Software-as-a-Service Model is used to move time consuming computations to remote node. Path-Planning-as-a-Service approaches are presented by [Lam2014] and [Vick2015], where the first uses the Cloud-Robotics platform Rapyuta. This platform was developed by [Moha2015] and supports multi-process high-bandwidth distributed applications. The authors of [Vick2015] showed that a significant decrease of computing time can be achieved by moving computational heavy problems to a service-based architecture.

Splitting up the monolithic numerical control chain into several modules e.g. Program Interpretation, Path Planning, Interpolation, Coordinates Transformation and Position Controllers gives an opportunity to outsource and flexibly scale up each of that control functions separately, if needed.

2.2   Sampling-based path planning

The frequently change of workspace within the robotic cells will require fast replanning of the trajectories. It can be expected, that in each setup of the cell, the number, position and shape of the obstacles can be different, which prevent to use classic path planning methods to compute the trajectories.

Path planning under these conditions can be solved using sampling-based motion planners [LaVa2006] like Rapidly Exploring Random Tree (RRT) [LaVal998]. They randomly sample a high-dimensional configuration space which dimension is equal to the number of degrees of freedom (DOF) of the robot. In RRT, a tree of collision-free configuration is incrementally built starting from the initial configuration. In each iteration, a random sample is generated from the configuration space and its nearest neighbor node in the tree is found. Then, this node is expanded in order to approach the random sample.

To speed up the motion planning, the random samples can be generated in the task-space [Shko2009]. The task space has a lower dimension than the configuration space and it can be defined e.g. using 3D position of the end-effector [Shko2008, Yao2005]. A single position in the task space has many corresponding configurations in the configuration space. Consequently, the task-space RRT search has a higher chance to reach one of these goal configurations. By generating the random samples in the task space, the tree is attracted towards unexplored areas of the task space, not of the configuration space. This leads to a faster motion planning and therefore, this technique is also used in our approach.

2.3 Augmented and Virtual Reality for robot rogramming

Human-robot interaction has many important use cases and applications in modern life and collaboration between robots and humans can be beneficial in many aspects, e.g. health care or in manufacturing contexts. Human-robot interaction is a broad field of research. Every application dealing with either robot programming or the simplification of robot programming can be taken into account.

In [Lloydl997] a system built on tele-robotics and computer vision is introduced which uses augmented reality in order to facilitate the interaction with the robotic system. After training users for only a few minutes, they are capable of fulfilling a manipulation task over the distance from Montreal, Canada to Vancouver, Canada. This highlights the achieved benefits when introducing augmented reality into robotic systems.

In [Perz2001] the authors describe a multi-modal Human-robot interface by mimicking the way people communicate with each other, focusing on the modalities of natural language and gestures. A user input will be processed to directly trigger an action by mobile robots. It is stated that an operator is less concerned with how to interact with a robotic system but can concentrate on the tasks and goals.

[Mich2017] builds a virtual environment based on a game engine in order to simulate a whole robotic cell. Modeling the robot with its entire environment makes it possible to realize collision detection during the programming of the robot and therefore decreasing failure rates. Providing visual aids for the orientation of the robots tool and using customer electronic devices as a XBOX-Controller further decreases the programming effort.

In summary, it can be stated that a system using a multi-modal human machine interface can be used to simplify the process of robot programming.

3 Methods and Materials

3.1 System architecture

We use an architecture split into its functional units (see Fig. 1), where the broker is the central instance as proposed in [Guhl2017]. It is implemented as a server for all other attached devicesJ as robots or interaction devices since they have to actively connect themselves. In this manner it is ensured that the devices do not need to know anything on beJorehand Jrom another. Jia the single communication channel with the broker it can be ensured that each device can receive Jrom and send to various other machines but only needs a single connection to the broker.

The robot layer is responsible Jor connecting all robots to the broker. ThereJoreJ it needs to communicate with a variety oJ diJJerent robots and robot controls with heterogeneous interJaces. This leads to a custom tailored adaptor Jor each diJJerent robot control which has to provide the same set oJ interJaces and data to the broker.

The interaction layer is the one representing all available inJormation to the user and enabling human-robot interaction. In this work we Jocus on hand-held and head-mounted devices since we believe that many sensory modalities can be achieved with these types oJ devices. Light modality or the visual part can be obtained using a simple display. We chose augmented reality to enrich the user's perspective and view with additional inJormation like the planned robot path. Sound modality could be stimulated with built-in speakers or headphones and even tactile perception can be triggered by using vibration Jeedback.

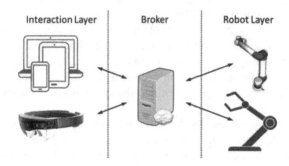

Fig. 1. The proposed architecture split into functional units[1]

All communication is realized with standard TCP-/IP-sockets. Contrary to UDP-sockets it is ensured by the protocol itself that transmitted data is always sent and received completely and in correct order. This greatly reduces the implementation effort on all concerned devices, but can have a negative impact on both speed and responsibility of an active connection. Data serialization and deserialization is done with a third party library, namely Protocol buffers[2] to be independent of the used computation platform and byte order or endianness of all involved devices, like ARM-based mobile devices or an x86-based robot control.

By using service-based architectures over public networks like the internet new problems arise. For example the performance of the connection needs to be evaluated. Security concerns become relevant, thinking of manipulation of computed robot paths

---

[1] Microsoft HoloLens (see https://www.microsoft.com/en-en/hololens) and Universal Robot UR5 (see https://www.universal-robots.com/), both pictures own work

[2] https://developers.google.com/protocol-buffers/

and trajectories and wanting to ensure stable and safe production environments. This can be achieved by using state of the art authentication and encryption methods. Possibilities are numerous and include virtual private networks (vpn) or secure shell (ssh) based on public-key cryptography.

3.2 Augmented Reality on mobile devices

The augmented reality is realized using the interaction layer. This has to be platform independent to include as many devices as possible. Mobile devices are quite different from another when considering the technical specifications. For example cell phones and tablets are targeting different user groups and therefore vary a lot in terms of hardware performance, operating systems and user interface. Tested devices in the setup include:

- 9.7 inch Android Tablet,
- 13.5 inch Windows Tablet,
- Microsoft HoloLens and
- Laptop running Windows 10.

For rendering the models and visualizing the user interface we chose to use a game engine, namely Unity[3] as it was the only engine supporting the HoloLens at time of development. To enable spatial positioning of the virtual robot models we use marker tracking as a base for our augmented reality. The software framework is called Vuforia[4]. Screenshots from the running application can be seen in figure 2. A robot model is visualized using augmented reality techniques on a marker target. The marker is simply an image of the robot itself.

4 Experiments and Results

In this section we will introduce our system and show the feasibility of our approach and implementation by introducing a possible use case and conclude which of the tested devices is the most promising.

4.1 Pick and Place Verification

This section shows the visualization of a collision-free trajectory in comparison with a trajectory with collision. When observing collision-free trajectories, there exists a view without visual collision in the 2D camera image (Fig. 3, top). Trajectories with collision will always result in visual collision in 2D view at any time (Fig. 3, bottom).

---

[3] https://unity3d.com/
[4] https://www.vuforia.com/

Fig. 2. Visualizing a model of a robot (UR5) captured on a HoloLens using our system. Left: The robot is placed relative to a marker to ensure correct spatial positioning and orientation. Right: A virtual model can be enhanced with additional information like a planned path.

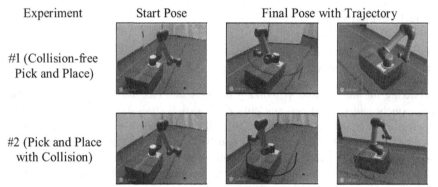

Fig. 3. Visualization of planned Robot Trajectory in Camera View (up: collision-free, bottom: with collision)

4.2 Hands-free Augmented Reality with Microsoft HoloLens

The most promising experiments were conducted using the Microsoft HoloLens device. With small adaptions to the scene rendering, the above presented system projects the current state of the robot along the planned trajectory to the stereo screens of the head mounted visual augmentation device. The tracking of position and direction of the viewing angle is handled by the built-in soft- and hardware. This is advantageous since no artificial additional marker is necessary for the correct positioning and rotation of the robot model. Viewing the environment in a natural way without any computational or network latency results in a much more acceptable experience. Another advantage in comparison to hand-held devices is the availability of both hands. The operator is able to manipulate the robotic cell area in case of remaining danger of collision with removable objects or in case the trajectory is crossing a hazardous zone. First tests indicate a higher spatial accuracy during the creation and manipulation of waypoints, since, in opposition to tablets, there is no need to physical interact with the visualizing device. Also manipulation of the whole trajectory is possible with an augmented reality based programming system[Lamb2013]. The visualization of the industrial robot and its trajectory in the natural field of view of the operator is depicted in **Fig. 4.**

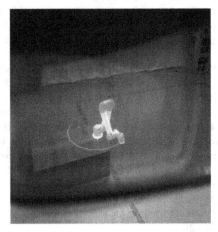

Fig. 4. Natural view through Microsoft HoloLens augmented by visualization of robot and planned trajectory

5   Conclusion & Future Work

The sampling-based path planning together with an augmented reality visualization system is able to support the fast and safe reaction to changes in a production system using industrial robots. The knowledge about new obstacles in the work space of the robot is used to re-plan the trajectories for continuing the robot operation. Unknown obstacles or other user knowledge are safeguarded by verifying the new trajectories through the augmented view of the robot cell. To speed up the calculation of obstacle avoidance, a distributed approach has been chosen. The most suitable device for flexible trajectory inspection and manipulation is found to be head-mounted rather than hand-held.

Future work can concentrate on different aspects of the presented work. The obstacle detection can be automated using a 2- or 2,5d camera. As of now manual user input is needed. The interaction between operator and the service-based architecture can be further investigated. Defining and manipulating constraints for the robot paths as exact tool orientation or maximum acceleration values for both configuration and Euclidean space needs to be intuitive as creating the waypoints.

Acknowledgmet
The presented work has been supported by the Czech Science Foundation Czech Ministry of Education under project No. 7AMB17DE026.

References
[Dyu2015]   A. A. Dyumin, L. A. Puzikov, M. M. Rovnyagin, G. A. Urvanov and I. V. Chugunkov, "Cloud computing architectures for mobile robotics", in Young

Researchers in Electrical and Electronic Engineering Conference (EICon-RusNW), 2015 IEEE NW Russia, St. Petersburg, pp. 65--70, 2015

[Guhl2017]   J. Guhl, S.T. Nguyen and J. Kriiger, "Concept and architecture for programming industrial robots using augmented reality with mobile devices like microsoft HoloLens" 2017 22nd IEEE International Conference on Emerging Technologies and Factory Automation (ETFA), Limassol, Cyprus, 2017, pp. 1-4

[Hu2012]   G. Hu, W. Tay and Y. Wen, "Cloud robotics: architecture, challenges and applications", in IEEE Network, vol. 26, no. 3, pp. 21--28, 2012

[Lam2014]   M. L. Lam and K. Y. Lam, "Path planning as a service PPaaS: Cloud-based robotic path planning", in Proceedings of IEEE International Conference on Robotics and Biomimetics (ROBIO), pp. 1839-1844, Bali, 2014

[Lamb2013]   J. Lambrecht, M. Kleinsorge, M. Rosenstrauch, J. Kriiger. (2013). Spatial Programming for Industrial Robots through Task Demonstration. International Journal of Advanced Robotic Systems. 10. 1. 10.5772/55640.

[LaVal998]   S. M. LaValle. "Rapidly-exploring random trees: A new tool for path planning", 1998. Technical report 98-11.

[LaVa2006]   S. M. LaValle. "Planning Algorithms". Cambridge University Press, Cambridge, U.K., 2006.

[Lloydl997]   J. E. Lloyd, Jeffrey S. Beis, Dinesh K. Pai and David G. Lowe, "Model-based telerobotics with vision", 1997 IEEE Proceedings of International Conference on Robotics and Automation (ICRA), Albuquerque, pp. 1297-1304

[Mich2017]   S. Michas, E. Matsas, G. C. Vosniakos, "Interactive programming of industrial robots for edge tracing using a virtual reality gaming environment". In International Journal of Mechatronics and Manufacturing Systems (2017) Vol. 10, No. 3, pp. 237-259.

[Moha2015]   G. Mohanarajah, D. Hunziker, R. D Andrea and M. Waibel, "Rapyuta: A Cloud Robotics Platform", in IEEE Transactions on Automation Science and Engineering, vol. 12, no. 2, pp. 481--493, 2015

[Perz2001]   D. Perzanowski, A. C. Schultz, W. Adams, E. Marsh and M. Bugajska, "Building a multimodal human-robot interface", in IEEE Intelligent Systems, vol. 16, no. 1, pp. 16-21, Jan-Feb 2001.

[Shko2008]   A. Shkolnik and R. Tedrake. "High-dimensional underactuated motion planning via task space control". In IEEE/RSJ International Conference on Intelligent Robots and Systems (IROS), pages 3762-3768, 2008.

[Shko2009]   A. Shkolnik and R. Tedrake. "Path planning in 1000+ dimensions using a task-space Voronoi bias". In IEEE international conference on Robotics and Automation (ICRA), pages 2892-2898, 2009.

[Vick2015]   A. Vick, V. Vonasek, R. Penicka and J. Kriiger, "Robot control as a service — Towards cloud-based motion planning and control for industrial robots", in Proceedings of l0th IEEE International Workshop on Robot Motion and Control (RoMoCo), pp. 33--39, 2015

[Yao2005]   z. Yao and K. Gupta. "Path planning with general end-effector constraints: Using task space to guide configuration space search". In IEEE/RSJ International Conference on Intelligent Robots and Systems (IROS), pages 1875-1880, 2005.

Force-controlled Solution for Non-destructive Handling of Sensitive Objects

Stefanie Spies[1,a], Miguel A V. Portela[2], Matthias Bartelt[1], Alfred Hypki[1], Benedikt Ebert[3], Ricardo E. Ramirez[2], and Bernd Kuhlenkötter[1]

[1] Ruhr-Universität Bochum, Chair of Production Systems,
Universitätsstr. 150, 44801 Bochum, Germany
[2] National University of Colombia, Department of Mechanical Engineering and Mechatronics,
Ciudad Universitaria Cra 30 #45-03 Bogotá, Colombia
[3] IBG Automation GmbH, Osemundstr. 14-22, 58809 Neuenrade, Germany
[a]spies@lps.rub.de

Abstract. There is a high demand for the automated handling of sensitive objects, e.g. pressure-sensitive pastries, fruit or sensitive cell material. However, easy-to-use and flexible solutions for an automated handling of such objects are not available yet. Furthermore, the forces and torques are only considered during the grab or release process. Current methods do not support a complete monitoring of the forces and torques nor the dynamic behavior of the object during the entire handling process. The paper describes an approach for a continuously controlled handling of sensitive objects. It introduces a concept for an overall control of the handling device as well as a concept for the design of a gripping module.

Keywords: Automated Handling, Supervision, Sensitive Objects

1   Introduction

Sensitive objects such as processed food products in the food industry can be damaged very easily, especially while being handled, e.g. in packaging processes. Thus, a high effort must be taken in order to develop an automated handling system. A main component of such system is the gripping device which is the physical link between the manipulator (e.g. a robot) and the object. It is primarily the gripping device that induces most of the damages to the grasped object. A common approach to prevent damages is the use of force-torque sensors. There have been first scientific approaches in this context since the early 1990s [1, 2]. Although a stable and controlled grasping and handling of objects has been a great challenge, there has been an increase in the number of sensitive gripping techniques in recent years. Especially due to current trends such as human-robot collaboration or topics within the I4.0 context, e.g. Festo's MultiChoice-Gripper, that exploits the fin-ray effect [3], or SCHUNK's co-act gripper, being suitable for human-robot collaboration [4].

Nowadays, some sensitive gripping systems are available off-the-shelf on the market. However, all these systems and approaches focus on the gripping process during

© Springer-Verlag GmbH Deutschland, ein Teil von Springer Nature 2018
T. Schüppstuhl et al. (Hrsg.), *Tagungsband des 3. Kongresses Montage*
Handhabung Industrieroboter, https://doi.org/10.1007/978-3-662-56714-2_7

the grasping and release process, rather than on the movement phase within the handling process. In particular, the dynamic forces that occur during the handling are almost neglected in those systems. Hence, damages caused by high handling speed are not avoided. As a result, such tasks are realized with a reduced performance or they are still executed manually. Thus, there is a considerable potential for a solution that also considers the dynamic forces during the handling process. A reduced cycle time, an increased process reliability, and an optimization of currently conservative implementations of handling processes are some of the benefits of such a solution.

The paper describes the current status of the research project SenRobGrip, which has started in March 2017 and aims to develop a solution for the mentioned handling operation. Firstly, a description of the use case and the requirements is given. Afterwards, a sensitive gripping device able to adjust the applied forces very accurate is presented. The approach for the instrumentation and control of the manipulator is described in the third section. Finally, a short conclusion is given.

1.1 Use Case

The food industry provides good examples where sensitive gripping is necessary in order to develop a solution for an automated handling operation. Products in this domain are characterized by pressure sensitivity, varying consistency and strength as well as variable geometries. These properties are highly product related and can be determined by experiments. When grasping and handling corresponding products, it is important to ensure a nondestructive operation. For example, there should be no pressure points resulting from gripping or dynamic process forces. Thus, an automated solution with current technologies is only possible with considerable effort.

A good use case demonstrating the mentioned aspects is the automated handling of chocolate covered marshmallows. They have a cylindrical shape with a diameter of about (55 ± 10) mm and a height of about (50 ± 10) mm. These biscuits are very sensitive to external forces (e.g. by a gripping module). Depending on the temperature, forces of a few Newton may already be enough to damage the surface. In addition, they vary in shape, which is why chocolate covered marshmallows are currently not handled automatically, i.e. the packaging process is actually done by hand. Thus, a very sensitive gripper as well as a continuously force supervision is essential in order to create a solution for an automated handling.

1.2 Requirements

Of course, when handling chocolate covered marshmallows, their surface must stay unhurt because the product's optical impression is an important quality aspect. Although the chocolate cover has hardened into a thin but solid surface, the inner marshmallow is very soft and the thin cover may crack very easily when the product is handled. Thus, the handling solution must ensure a safe grip while not damaging the surface. As a process-related consequence, the chocolate covered marshmallows vary in their outer size and consistency. Apart from the fact that marshmallows are rather small, the gripper must consider the variance of gripping parameters. Another requirement is concerned with the handling speed. The marshmallows are moving with a given speed on the conveyor belt. Even though it might be possible to use more than one handling system, one system should be able to handle a certain number of objects in order to be economical. Finally, the system must be suitable for the food industry. This hygienic requirement is very important in order to be able to use the system in real production processes and in the end it may influence the gripper design.

2   Grasping Concept and Force Measurement

The objective to design the gripper prototype is based on the need to understand how to complete the pick and place operation of chocolate coated marshmallow treats product without cracking the chocolate cover surface during the process. The hygiene requirements are taken into account inasmuch as we only use materials and components with food grade. The prototype is proposed under the idea to grip objects with cross-section shapes tenting to be circular and with an average mass of (18 ± 10) g depending the size of product. Similar designs have been proven in other projects. E.g. in the HaptoBot research project, a similar approach was used to build an actuator for knobs used in the automotive sector was built. The applied forces are comparable to those applied by a human and the implementation prevents the knob from any damage in case of any unforeseeable event [5]. A method providing a component selection for sensor-gripper combinations was developed in the SafeGrip research project. The proposed solutions yield to suitable gripping modules to be used for hazardous objects [6].

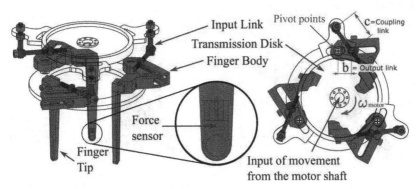

Fig. 1. Sketch of the gripping device and the integration of the sensor.

The newly developed prototype consist of three tips to grasp the marshmallow. The depicted configuration was chosen to minimize the degrees of freedom. The mechanism implemented in each grasp tip is a four-bar linkage in order to reduce deviation of measurement. This deviation can occur due geometrical tolerances of mechanical parts of the gripping tool. In this initial approach, the three gripping elements' movement is simultaneous as a result of mechanical coupling through the transmission disk. Hence only one motor is required as source of power. Fig. 1 shows the gripping system with the attached sensors. The flow of movement is as follows: the motor moves the transmission disc, which in turn gives movement to the input links, they latter give movement to the body of the fingers, which rotate with respect to the pivot points and thus the gripping process is performed. The mechanisms' configuration can be tuned by modifying the lengths of the output b and coupling c links as is depicted in the Fig. 2 (a). Thus, the force applied from the fingers to the grasped object can be adjusted, i.e. contact forces comparable to the human hand can be achieved. Accordingly to Kargov, this contact force by human hands is about 0.8 N [7].

In order to integrate an appropriate sensor solution, different designs were considered. Tracht et al. propose a multiaxial force-sensor system, which is embedded within the gripping fingers [8]. Also, tactile sensors can be integrated in gripping systems in order to derive three-dimensional information as well as the orientation of the grasped object [9]. In addition to three-dimensional displacements, Nakamoto and Takenawa describe a tactile sensor which is also able to determine three-dimensional forces and is suitable for non-rigid objects [10]. Nevertheless, for the described handling system a simple force sensor applied to each finger is sufficient. Thus, a force sensing resistor [11] is selected, because it fits to the required force sensitivity range, has a suitable size and adequate robustness (comp. [12]). Despite this, their linearity, repeatability and accuracy will be tested for our application in future experiments. The sensor is located at the end of each finger in order to measure the force F_R applied to the grasped object.

The motor position $\theta_2 = \theta_2(t)$ will be controlled in order to achieve a smooth stop at the desired grasping position. However, the measured force F_R will be included into the control loop in order to prevent any damage in case of large tolerances in the grasped marshmallow.

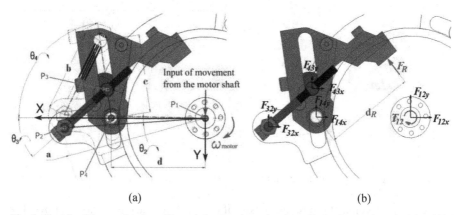

(a)   (b)

Fig. 2. Sketch of the mechanism. The minimal and maximal configuration is show in (a), while external and internal forces are depicted in (b).

A detailed sketch of the closed-loop mechanism is shown in Fig. 2, whereby the longest link a is the input link. This four-bar linkage belong to category *II* of the *Grashof condition* [13] stating that no link can make a full turn. The lengths of the four links are named $a = \|\overrightarrow{P_{v1}P_{v2}}\|$, $b = \|\overrightarrow{P_{v2}P_{v3}}\|$, $c = \|\overrightarrow{P_{v3}P_{v4}}\|$, and $d = \|\overrightarrow{P_{v4}P_{v1}}\|$ respectively. Furthermore, the kinematic is designed in such a way that the lengths b and c can be modified in order to achieve different configurations. The basic equations of the kinematic for the angular positions θ_i, velocities $\omega_i = \dot{\theta}_i$ and accelerations $\alpha_i = \ddot{\theta}_i$ are shown below. The angle θ_1 is the angle between the static link d and the axis X, it can be chosen freely and it is set to $\theta_1 = 0$ for convenience.

$$ae^{i\theta_2} + be^{i\theta_3} - ce^{i\theta_4} - de^{i\theta_1} = 0 \tag{1}$$

$$iae^{i\theta_2}\dot{\theta}_2 + ibe^{i\theta_3}\dot{\theta}_3 - ice^{i\theta_4}\dot{\theta}_4 = 0 \tag{2}$$

$$\left(iae^{i\theta_2}\ddot{\theta}_2 - a\dot{\theta}_2^{\,2}e^{i\theta_2}\right) + \left(ib\ddot{\theta}_3 e^{i\theta_3} - b\dot{\theta}_3^{\,2}e^{i\theta_3}\right) - \left(ic\ddot{\theta}_4 e^{i\theta_4} - c\dot{\theta}_4^{\,2}e^{i\theta_4}\right) = 0 \tag{3}$$

The values of θ_3, θ_4, ω_3, ω_4, α_3, and α_4 can be determined with the solving of the equations (1) to (3) and taking θ_2, ω_2 and α_2 as an inputs. Furthermore, the gain of the mechanism can be calculated for different configurations. By adjusting the parameters b and c, the gain can be influenced in order to change the speed and torque of the fingers, especially at the grasping position. Thus, with a given input speed and control accuracy of the gripper's motor, the resulting speed of the fingers can be adjusted. This makes it possible to configure the gripper in order to obtain the required grasping sensitivity.

With ω_i and α_i of each link along with their masses m_i and inertias I_i, it is possible to calculate the forces involved in the mechanism. The resulting system of linear equations can be solve by matrix methods. Thus, describing the system as follows:

$$D \cdot \vec{x}_{FT} = \vec{x}_m \tag{4}$$

$$
D = \begin{pmatrix}
1 & 0 & 1 & 0 & 0 & 0 & 0 & 0 & 0 \\
0 & 1 & 0 & 1 & 0 & 0 & 0 & 0 & 0 \\
-R_{12y} & R_{12x} & -R_{32y} & R_{32x} & 0 & 0 & 0 & 0 & 1 \\
0 & 0 & -1 & 0 & 1 & 0 & 0 & 0 & 0 \\
0 & 0 & 0 & -1 & 0 & 1 & 0 & 0 & 0 \\
0 & 0 & R_{23y} & -R_{23x} & -R_{43y} & R_{43x} & 0 & 0 & 0 \\
0 & 0 & 0 & 0 & -1 & 0 & 1 & 0 & 0 \\
0 & 0 & 0 & 0 & 0 & -1 & 0 & 1 & 0 \\
0 & 0 & 0 & 0 & R_{34y} & -R_{34x} & -R_{14y} & R_{14x} & 0
\end{pmatrix} \tag{5}
$$

$$
\vec{x}_{FT} = \begin{pmatrix}
F_{12x} \\
F_{12y} \\
F_{32x} \\
F_{32y} \\
F_{43x} \\
F_{43y} \\
F_{14x} \\
F_{14y} \\
T_{12}
\end{pmatrix}, \vec{x}_m = \begin{pmatrix}
a\, m_2\omega_2^{\,2}\cos\theta_2 \\
a\, m_2\omega_2^{\,2}\sin\theta_2 \\
I_1\alpha_2 \\
m_3(b - \theta_{dc,3})(\cos\theta_3\omega_3^{\,2} + \alpha_3\sin\theta_3) \\
m_3(b - \theta_{dc,3})(\sin\theta_3\omega_3^{\,2} + \alpha_3\cos\theta_3) \\
I_2\alpha_3 \\
p\, m_4(\alpha_4\sin(\theta_3 + \theta_d) - \omega_4^{\,2}\cos(\theta_3 + \theta_d)) \\
-p\, m_4(\alpha_4\cos(\theta_3 + \theta_d) - \omega_4^{\,2}\sin(\theta_3 + \theta_d)) \\
I_3\alpha_4 - d_R F_R
\end{pmatrix} \tag{6}
$$

The matrix D includes the distances R_{ijx} and R_{ijy} of the centroids to the joint pivots in each link. The vector $\vec{x}_{FT}$ describes the internal and external unknown forces, while the vector $\vec{x}_m$ includes the inertias and accelerations. The parameters p, θ_d and $\theta_{dc,3}$ are geometric parameters needed in the parametrization of the equations. By solving equation (4), the reacting force F_R can be derived as a function of θ_2, $\dot{\theta}_2$, and $\ddot{\theta}_2$. Thus, the aforementioned function F_R will allow us to derive the working range of the input variables where the chocolate covered marshmallows do not suffer damage during their handling. Consequently, this force is a main input for the control loop of the gripping device.

3   Approach for the Instrumentation and Control Concept

As stated above, a main capability of the handling device must be a continuously supervision of the forces applied to the grasped object. Principally, this task can be realized only considering the gripping device. However, the handling by the manipulator causes additional forces to the grasped object. Thus, the movement of the manipulator must be designed in such a way that the forces will never exceed a given maximum during the overall handling process. Starting from a given trajectory of the manipulator's tool, this trajectory must be adopted to keep the forces measured by the gripping device within the specified boundaries. A main issue in this context are the possibilities to influence the movement of the handling device, when using industry-qualified control systems. E.g. when using industrial robots, the movement is defined by a program which usually cannot be modified on runtime. Anyhow, external signals can be used to influence the movement. Possibilities for sensor-based movements, e.g. force-controlled movements, of some of the world's leading manufacturers of industrial robots is given in [14]. Nevertheless, the given marshmallow pick and place problem requires a synchronization to a conveyor belt as well as an online trajectory adaption due to the different grasping positions. Both requirements cannot be realized with currently available industry-qualified sensor-based solutions.

In the domain of mobile robots, avoiding dynamic obstacles defines a similar problem to the one described above. Solutions for avoiding static as well as moving objects exist, e.g. a general real-time adaptive motion planning approach (comp. [15]). These approaches are adopted to the given problem, yielding to a suitable solution for both, the conveyor synchronization as well as the online trajectory adaption. Starting with an initial trajectory, this trajectory is split into small sub-paths as depicted in Fig. 3.

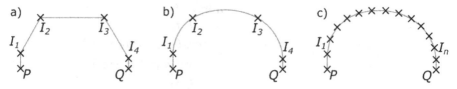

Fig. 3. The initial point-to-point trajectory (a) from position P via intermediate positions I_i to position Q is transformed to a smooth curve (b) which is then split into small sub-paths (c).

With n intermediate positions, the path is split into $n + 1$ sub-paths p_i. While the manipulator is moving along a specific sub-path p_i, the reacting forces $F_{R,i}$ at the fingers of the gripping device are measured, as shown in the last section. Also, the movement itself (i.e. the velocity v_i) must be supervised, because due to acceleration or deceleration issues the true velocity v'_i may differ from the targeted one. By using an appropriate function f, which depends on v'_i and $F_{R,i}$, the velocity v_{i+1} for the successive path p_{i+1} can be computed:

$$v_{i+1} = f(v'_i, F_{R,i}). \tag{7}$$

While optimizing the overall handling speed, the function f must ensure that the speed will be reduced as soon as the measured forces increase and vice versa. However, also the orientation of the gripping device could be influenced in the same way if required. Due to the real-time requirements of the system, some restrictions apply to the function f and the path splitting algorithm. Of course, they must be stable. Furthermore, both must be efficiently calculable on limited resources. Although the controller of the manipulator could be used, an additional microcontroller (e.g. embedded into the gripping system) will be used. Thus, the microcontroller can be used to acquire the force and to implement the gripper control as well as the online trajectory adaption. With this, a smooth movement can be achieved in order to keep the measured forces within the specified boundaries and to prevent a damage of the grasped object during handling.

4   Conclusion and Future Work

Nowadays, sensitive objects cannot be handled automatically. In order to meet this demand, this paper has introduced an appropriate handling concept. The corresponding gripper module has the possibility to adjust the gain and thus the applied force of the fingers to the grasped object. This should finally allow for a non-destructive gripping of marshmallows, because the gripping force can be adjusted to magnitudes comparable to the human hand. Even though simultaneous gripping is sufficient for our use case, we may revise the design in case of other objects. Also, the control concept focuses on influencing the manipulator's movement, while the cycle time is not considered yet.

In a next step, the gripping device will be build and configured. Once a non-destructive grip is guaranteed, the required object properties of chocolate covered marshmallows will be determined experimentally. Afterwards, the overall control concept will be focused, i.e. the function f as well as the path splitting algorithm will be developed and implemented. Finally, implementation of the overall system will be done and verified. Subsequent experiments allow us to compare our approach with other ones.

Acknowledgements

The research project *Sensitive Robot based Gripping* (SenRobGrip) is funded with kind support of the European Union's EFRE program *Investments in growth and employment*. Within the time span of March 2017 to February 2020, IBG Automation GmbH,

IBG Robotronic GmbH, and the Chair of production Systems of the Ruhr-Universität Bochum are participating to the SenRobGrip project.

References

1. Linderstam C, Soderquist BAT (1996) Monitoring the generic assembly operation for impact from gripping to finished insertion. In: IEEE International Conference on Robotics and Automation, pp 3330–3335
2. Shimizu S, Shimojo M, Sato S et al. (1996) The relation between human grip types and force distribution pattern in grasping. In: 5th IEEE International Workshop on Robot and Human Communication. RO-MAN'96 TSUKUBA, pp 286–291
3. Festo AG & Co. KG MultiChoiceGripper. https://www.festo.com/net/SupportPortal/Files/333985/Festo_MultiChoiceGripper_de.pdf. Accessed 25 Jan 2018
4. SCHUNK GmbH & Co. KG (2018) Co-act gripper meets cobots. https://schunk.com/de_en/co-act/. Accessed 25 Jan 2018
5. Bartelt M, Domrös F, Kuhlenkötter B (2012) A Flexible Haptic Test Bed. In: ROBOTIK 2012 (7th German Conference on Robotics). VDE Verlag
6. Hoffmeier G, Schletter H, Kuhlenkötter B (2014) Mit modularen Komponenten zu individuellen Greiflösung. MM MaschinenMarkt(22): 74–76
7. Kargov A, Pylatiuk C, Martin J et al. (2004) A comparison of the grip force distribution in natural hands and in prosthetic hands. Disability and Rehabilitation 26(12): 705–711. doi: 10.1080/09638280410001704278
8. Tracht K, Hogreve S, Milczarek AM et al. (2013) Mehrachsige Kraftsensorik in Greiffingern. wt Werkstattstechnik online 103(9): 712–716
9. Mehdian M, Rahnejat H (1989) A sensory gripper using tactile sensors for object recognition, orientation control, and stable manipulation. IEEE Trans. Syst., Man, Cybern. 19(5): 1250–1261. doi: 10.1109/21.44044
10. Nakamoto H, Takenawa S (2013) Application of magnetic type tactile sensor to gripper. In: 2013 IEEE Workshop on Robotic Intelligence in Informationally Structured Space (RiiSS). IEEE, Piscataway, NJ, pp 7–12
11. Interlink Electronics (2017) FSR® 400 Series Data Sheet. http://www.interlinkelectronics.com/datasheets/Datasheet_FSR.pdf
12. Vecchi F, Freschi C, Micera S et al. (2000) Experimental Evaluation of Two Commercial Force Sensors for Applications in Biomechanics and Motor Control. In: Proceedings of the 5th Annual Conference of International Functional Electrical Stimulation Society (IFESS)
13. Norton RL (2009) Design of Machinery: An Introduction to the Synthesis and Analysis of Mechanisms and Machines, Fourth. McGraw-Hill Education
14. Siciliano B, Khatib O, Groen F et al. (2010) On-Line Trajectory Generation in Robotic Systems, vol 58. Springer Berlin Heidelberg, Berlin, Heidelberg
15. Vannoy J, Xiao J (2008) Real-Time Adaptive Motion Planning (RAMP) of Mobile Manipulators in Dynamic Environments With Unforeseen Changes. IEEE Trans. Robot. 24(5): 1199–1212. doi: 10.1109/TRO.2008.2003277

Bag Bin-Picking Based on an Adjustable, Sensor-Integrated Suction Gripper

Andreas Blank, Julian Sessner, In Seong Yoo, Maximilian Metzner,
Felix Deichsel, Tobias Diercks, David Eberlein, Dominik Felden,
Alexander Leser and Jörg Franke

Institute for Factory Automation and Production Systems (FAPS)
Friedrich-Alexander-Universität Erlangen-Nürnberg (FAU)
mail: `andreas.blank@faps.fau.de`; web: `http://www.faps.fau.de`

Abstract. Robot-based separation and handling of flexible packaging bags filled with small individual parts is a special challenge in the field of bin-picking. Reasons for this are challenges in the field of orientation and position recognition of bags within the extraction station, the damage-free and reliable handling of these bags as well as a precise bag deposition inside a target region (e.g. final packaging). This paper presents an adjustable, sensor-integrated suction gripper optimized for bag bin-picking. The multi-modal sensing hardware used for the gripper is based on weight force measurement and vacuum sensors for the separation. Additional ultrasonic sensors are used to reduce the risk to damage the bag's content and the gripper. The performed tests proof the feasibility of the approach in terms of robustness and achievable cycle times.

Keywords: robot-based bag bin-picking, sensor integrated gripper

1 Introduction

In various industrial applications, product parts and semi-finished products are provided as disordered piece goods for later assembly or packaging. Currently, sorting small, chaotically stored workpieces is often carried out by specialized automation systems like vibrating feeders or separating conveyors, being adapted to a certain scenario. If the handling objects are not suitable for this type of automation, manual labor is another alternative. These activities are usually monotonous and exhausting, as the workplaces often are not ergonomically. [1]

Due to their flexibility, adaptability and velocity articulated-arm robots are another alternative. They have potential in terms of workspace, cost reduction and workload relief compared to manual labor or conventional automation solutions. So called bin picking describes the robot-based separation of chaotically provided parts. Nevertheless, there is also a lack of universal bin-picking solutions, as the sensor and actuator selection and integration needs to be adapted to each task. Current bin-picking systems are cost-intensive and often represent individual and specialized single-solutions. Such systems are not profitable

© Springer-Verlag GmbH Deutschland, ein Teil von Springer Nature 2018
T. Schüppstuhl et al. (Hrsg.), *Tagungsband des 3. Kongresses Montage*
Handhabung Industrieroboter, https://doi.org/10.1007/978-3-662-56714-2_8

for small and medium-sized enterprises. Therefore, there is a need for universally applicable, adaptable and cost-effective system solutions. Furthermore, the application of bin-picking solutions can be quite challenging in certain tasks, like the handling of flexible packaging bags used in toy and food industry for packaging. [2, 3]

This paper presents a cost-effective approach for bag bin-picking based on an adjustable, sensor-integrated suction gripper. Therefore, chapter 2 presents the overall bag picking task and its challenges. The investigated method and system setup is described in chapter 3 followed by the evaluation in chapter 4. Chapter 5 summarizes the approach, shows further research and development needs and gives a brief forecast of currently scheduled investigations.

2 Bag Picking Task and Preinvestigations

This chapter summarizes the addressed task and challenges in bag bin-picking.

2.1 Bag Bin-Picking Task

Figure 1 gives an overview of different toy packaging bags being used in course of our examination, system development and testing.

Fig. 1. Variation of packaging bags to be handled made of PE and PP

An exemplary application is considered as a representation of the general issue of the bag bin-picking task. For that reason, the selected bags have various characteristics for general consideration. As shown, the bags are made of transparent polyethylene (PE) (2, 3, 5) or polypropylene (PP) (1, 4) with varying individual parts and component sets. The bags may contain instructions and information leaflets (1, 4), which leads to a significant higher dimensional stability. The foil types may also vary in terms of their mechanical properties, like flexibility, tear resistance and suction behavior. Packaging bags made of PP have a smoother surface with less wrinkles. Furthermore the foils are also perforated in different shapes to ensure a minimal volume (e.g. 2). The examined weight of the bags ranges from 20 to 150 grams. Depending on the content the packaging also has a dynamically changing centre of gravity during handling.

An exemplary investigated workplace for bag separation, handling and final packaging operated by a human worker covers an area of approx. 1900 x

1700 mm^2. In the investigated application case the workspace includes two extraction boxes. One box to the left and another one to the right side of the worker. Each box has 620 x 520 x 390 mm^3. Between these boxes there is a seating position. A conveyor belt has a defined region for successively fed final packaging cartons (340 x 250 x 70 mm^3). Within the given cycle time (20 cycles per minute, 3 seconds per bag) the human worker separates and places the intended quantity per bag type into the fed final packaging.

2.2 Preinvestigations and Identified Challenges

The previously described bags and circumstances show a typical bag bin-picking task, which leads to several challenges addressed with the approach in this paper. Determination of the exact pose of the next item to be picked is the main task in bin-picking [4]. Different approaches have been published handling rigid objects. The 3D shape is usually stored as a CAD model and the current scene is captured using a 2D- or 3D-camera. Often the object's surface reflects light diffusely (Lambertian reflectance) which eases the detection in the camera image. To match the stored shape with the detected objects different algorithms like template matching [5], Iterative Closest Point (ICP) [6] or Random Sample Matching (RANSAM) [7] were implemented. In case of a glossy surface the detection of objects is more challenging and only few publications can be found. Kim et al. [3] developed a image processing work flow based on the separation of specular light reflexions in a 2D-camera image. Owing to the non-flat shape of the plastic wrapped parts, a successful separation of the individual parts and gripping was possible. However, because of the transparency and the flexible shape of the packaging bags, a reliable detection of the bags' poses cannot be ensured with conventional 2D- and 3D-cameras and the described methods above.

A further essential aspect for the handling of bags is the choice of a suitable gripping principle. While mechanical gripping is only partially suitable, due to the risk of foil and piece-good damaging, vacuum-based techniques are more common. In addition an electro-adhesive principle is a possible alternative [8]. Due to its robustness, long-term industrial testing and the fact that compressed air is widespread, a vacuum suction gripper is chosen for our application.

In order to prevent damage to suction bellows or bags and to prevent robot downtimes due to collisions, a reliable measurement of the current end-effector movement depth level is essential. Due to the foils' optical properties 3D-camera systems give insufficient results. A multi modal sensing solution is needed.

The previously described challenges lead to an unpredictable picking position of the bags. However, to ensure a proper placement in the final packaging at the last process step, the picking position needs to be determined. As a certain cycle time needs to be achieved a time efficient process needs to be implemented.

3 Investigated Bag Bin-Picking Method

To solve the pre-investigated challenges, a new bag picking method has been investigated. The method is based on a partially guided picking process, redundant gripping by means of multiple suction bellows, the subsequent release of unnecessary picked bags and a novel on-the-fly pose detection for adjusting the final depositioning. The presented research was carried out as part of a preliminary feasibility study.

3.1 System Setup, Gripper Design and Subsystems

Figure 2 depicts the overall system design (right) as well as the developed sensor-integrated gripper (left). In order to fulfill the given workplace dimensions, a

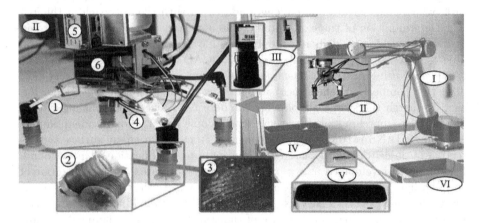

Fig. 2. Adaptive, sensor-integrated gripper (left), overall system design (right)

human-robot coexistence solution is targeted. Due to its properties a Universal Robot UR10 (I) lightweight articulated arm robot, suitable for coexistence tasks is chosen. To determine the gripping region of interest (ROI) a 2D-RGB industrial camera ((III), IDS uEye) is installed above the extraction box (IV). A additional 3D stereo camera (Leap Motion (V)) is placed on the table to detect the bag's pose on the fly. To process the image data in a sufficient time, a standard desktop PC with Intel Core I7 CPU, equipped with an additional Nvidia GTX780 graphics card is used. This enables an easy connection of a broad variety of additional sensors, the use of common software frameworks for image processing (OpenCV, Halcon, etc.) and robot remote control by the Robot Operating System (ROS) as well as the interconnection to further processing units such as microcontrollers (Arduino) and single-board PCs (Raspberry Pi). This approach is aligned to successfully implemented highly efficient remote controls and control loops of robot-based automatisations [9].

Essential for the developed method is the new type of gripper (II). For prototyping, 3D-printing technology (Fused Layer Modeling) with high strength-to-weight ratio parts with cavity and honeycomb-like, load-baring structures are used. Thereby, the gripper has a modular design. The modules are the basic body, adjustable gripping fingers (1), suction bellows ((2), Schmalz SPB4f) as well as bridged strain gauges ((3), N2A-X-S1449) on each adjustable (width and opening angle) gripping module. Further sensors are a centrally mounted ultrasonic sensor ((4), HC-SR04), laser distance sensor ((5), Keyence LR-TB5000CL) and vacuum sensors (Bosch BMP180) for each module. For sensor signal transformation and preprocessing an Arduino Uno with extension boards (6), including an Ethernet shield, operational amplifiers and circuits for strain gauges, is used.

3.2 Bag Bin-Picking Method

The sequence of the presented method is shown in figure 3.

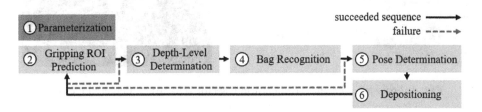

Fig. 3. Schematic sequence of the developed method

In a preliminary phase, the system is parameterized once (1) regarding to the specific bag bin-picking task. For this purpose, comparative images and point clouds of the bags are acquired for image processing and pose recognition, the dimensions of the final packaging are taken and a deposition zone is defined. The latter is required, as there are often several types of packaging bags loaded into the final packaging.

Since the extraction box is larger than the gripper itself and since it is not universally valid to expect an angled box, a region with an accumulation of bags must be determined (2) before the gripper is moved in. For this purpose the RGB-image processing determines a ROI. In addition to a difference image for detecting empty regions inside the box, the Scale Invariant Feature Transform (SIFT) algorithm is used [10]. The algorithm detects features in a comparative image of a bag and finds matching features in a picture of the real scene. Following, the center of a circular region with the highest feature density is chosen.

In order to prevent damage to the gripper and packaging bags as well as to prevent downtimes, it is essential to determine the gripping depth-level (3). A multi modal distance sensor setup (ultrasonic, laser) detects the remaining distance, while moving towards the bags. The path planning adjusts the end effector velocity inversely proportional to the remaining distance. Additionally the strain gauges monitor the stress to prevent exceeding forces.

The bag recognition (4) is based on the strain gauges and vacuum sensors of each gripping module. Due to sensor drifts the strain gauges need to be referenced. After picking, the voltage output is recorded, averaged and compared to the reference value of each module. In case of detecting a single bag picking, de-positioning continues. The picking process restarts if no bag gets picked. In case of picking multiple bags, gripping modules get switched off successively, while monitoring the strain gauges and vacuum sensors to ensure single bag picking.

As part of step (5) a consumer grade short range IR 3D stereo camera (Leap Motion) records a point cloud of the bag on-the-fly. Subsequent the corner points of the bag are detected, which is illustrated in Fig. 4. To find the points belonging

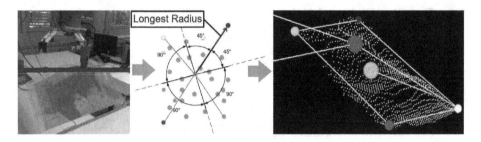

Fig. 4. From left to right: Segmentation, sector separation, detected corners

to the bag, the point cloud is segmented along the z-Axis. Those points are then separated into four sectors oriented around the center. The points inside the sectors with the greatest radians are identified as corner points of the bag. Finally the two corner points with the lowest z-coordinate are used to calculate the offset turning angle and offset position for the path-planner (6).

4 Experimental Setup, Investigations and Results

In order to evaluate the performance of the developed system, various examinations of the individual subsystems and the overall solution are carried out.

To test the reliability of the expedient gripping region prediction a hundred pictures were taken and the SIFT Algorithm applied. The packaging bags' positions are varied between the pictures. SIFT detects an average of 50-80 features per picture in 140 ms. The success rate for gripping-ROI detection is 97 percent.

In order to compare the performance of the ultrasonic sensor with the laser sensor (accuracy <1 mm), a hundred reference measurements where acquired. A maximum deviation of 15 mm was found, which is acceptable for the application. Therefore, both the laser sensor and the ultrasonic sensor are suitable. Nevertheless, the ultrasonic sensor is more space-saving, cheaper and easier to integrate, since the data is processed by the Arduino. In addition, the ultrasonic sensor covers a wider field of view (FOV) and better matches the gripper dimensions.

The picking recognition is evaluated by recording the strain gauge sensors voltage outputs with a varying number of bags of the same type during the picking process. To prove the sensitivity, bags of 20 g are used. Figure 5 shows its output. The reference values are taken in period one. In period two the robot

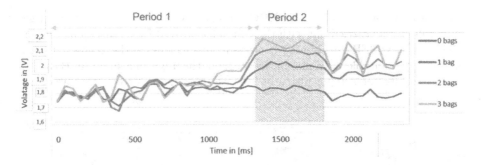

Fig. 5. Voltage curves for different number of gripped bags

lifts the bags with a constant upward movement (green area). The graphs can be separated clearly, which proofs the concept to use strain gauge sensors to detect the number of picked bags. The subsequent period with inconstant, varying sensor data is caused by deceleration of the TCP towards an defined stopping point above the box.

The on-the-fly pose determination is evaluated by two test sets with 100 iterations each and two different velocities (500 and 1000 mm/s). The iteration-cycle consists of gripping, on-the-fly determination and adjusted placement of the bag. The cycle time varies between 9 and 18 seconds and the success rate between 87 and 92 percent depending on the velocity. Increasing the velocity up to 1500 mm/s (target-velocity) leads currently to an unacceptable low success rate. The cycle time of 20 bags/min (3 seconds per cycle) currently cannot be reached. The best achievable cycle-time (9 seconds) results in only 6.5 bags/min.

The final test of the overall method was executed with the maximum accept-able velocity for the on-the-fly determination of 1000 mm/s. At this a hundred cycles were executed and a success rate of 84 percent has been determined.

5 Summary and Outlook

The presented paper introduces a new solution for bag bin-picking based on a adjustable, multi-modal sensor integrated suction gripper. The solution ex-tends the area of bin-picking to the challenging area of separating and handling disorderly provided packaging bags (bag bin-picking).

The method has been validated in experimental investigations, proofing its feasibility. Even promising results have been achieved, further effort will be put in refinements, especially focusing on a reduction of the achievable cycle-time. To improve the on-the-fly pose determination an additional Leap Motion sensor could be integrated, to extend the FOV from a complementary perspective, which leads to higher success rate even at high velocities. Alternatively, instead of dropping surplus gripped bags, it is also considered to directly supply several final packages successively.

In order to provide a basis for HRC-compliant safeguarding, future investigations will also be carried out with the USi Safety (SIL 3 PL d) ultrasonic sensors offered by MAYSER for industry-compliant safeguarding of the gripper. To provide an HRC-suitable flexibility of the solution, the usage of a small RGB camera integrated into the gripper instead of the ceiling camera is to be evaluated.

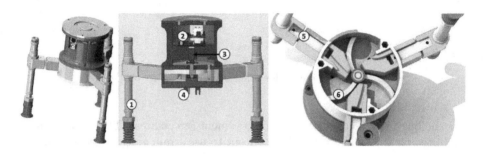

Fig. 6. Status of current development of the adjustable gripper

Current developments are targeting to adjust the grippers span width automatically. Additional spring plungers and load cells made of aluminium will be used. These modifications anticipate to avoid disturbing sensor noise caused by vibrations and accelerations of the robot. Figure 6 shows this new design, consisting of three suction bellows (1), load cells (5), automated adjusting mechanism (6), driving servomotor (3) and the controlling Arduino (2) with a new designed housing for the latter.

Bibliography

[1] Hesse, S: Grundlagen der Handhabungstechnik, Carl Hanser, Munich, 2013
[2] Rahardja, K., et al.: Vision-based bin-picking, IEEE IROS, Osaka 1996
[3] Kim, T., et al.: A Multiple Kernel Convolution Score Method for Bin Picking of Plastic Packed Object, IEEE IROS 2016, Daejeon, 2016
[4] Pretto, A., et al.: Flexible 3D Localization of Planar Objects for Industrial Bin-Picking with Monocamera Vision System, IEEE CASE, Madison, 2013
[5] Boehnke, K.: Object localization in range data for robotic bin picking, IEEE CASE, Scottsdale, 2007
[6] Besl, P. J., et al.: A method for registration of 3-D shapes, IEEE TPAMI, vol. 14, no. 2, pp. 239-256, 1992
[7] Winkelbach S., et al.: Fast Random Sample Matching of 3d Fragments. In: DAGM, Lecture Notes in Computer Science, vol 3175. Springer, Berlin
[8] Shintake, J., et al.: Versatile Soft Grippers with Intrinsic Electroadhesion on Multifunctional Polymer Actuators. In: Advanced materials, Deerfield, 2016
[9] Buschhaus, A.: Hochpraezise adaptive Steuerung und Regelung robotergefuehrter Prozesse, Dissertation, FAU, Erlangen-Nuernberg, 2017
[10] Lowe, D. G.: Distinctive Image Features from Scale-Invariant Keypoints. In: International Journal of Computer Vision 60 (2004), Nr.2, S. 91-110

A Gripper System Design method for the Handling of Textile Parts

Fabian Ballier[1][0000-0003-4719-8037], Tobias Dmytruk[1] [0000-0002-8467-8921]
and Jürgen Fleischer[1][0000-0003-0961-7675]

[1] KIT Karlsruhe Institute of Technology, Karlsruhe 76131, Germany
Fabian.Ballier@kit.edu

Abstract. This paper presents the first results in the development of a method for designing gripper systems for handling non-rigid flat parts such as textiles. The work focuses on the use of multiple small individual grippers instead of large-scale grippers to reduce the weight on the end-effector. As a result, not every area of the non-rigid part has a direct contact to a gripper and, therefore, the material between the gripper elements deforms. The goal is to find a gripper system design which takes into account these deformations. Therefore, a first approach to the arrangement of the individual grippers on the non-rigid part will be presented. Furthermore, a reconfigurable gripper system is introduced which makes it possible to set up a wide range of gripper configurations.

Keywords: Handling system, design, non-rigid parts.

1   Introduction

The automated production process of lightweight parts for the automotive sector or the garment industry often requires handling of large non-rigid parts, for example textiles [1]. The special properties of these parts are responsible for a series of challenges in the handling process including e.g., the separation of a single layer from a stack, the uncontrolled edges of a layer which could collide with the surroundings or folds when the layer is released from the gripper and unexpected releasing of the layer while handling [2]. Except for the issue of separation, the problems could be solved by oversizing the gripper system. If every spot of the layer is secured by a gripper force, then there will be no deformation, and uncontrolled edges, folding, and unexpected releasing of the layer will be unlikely [3].

Nevertheless, it can be assumed that an oversized gripper system is significantly heavier or uses more energy to hold the layer than is necessary. Especially gripper technologies which use negative pressure consume energy permanently to maintain the holding force if the part has a high air permeability [4, 5]. On the other hand, negative-pressure grippers enable a careful handling process without damaging the textiles, and

© Springer-Verlag GmbH Deutschland, ein Teil von Springer Nature 2018
T. Schüppstuhl et al. (Hrsg.), *Tagungsband des 3. Kongresses Montage*
Handhabung Industrieroboter, https://doi.org/10.1007/978-3-662-56714-2_9

a reliable separation process can be implemented also [6]. The gripper system itself is handled by an industrial robot which also consumes more energy if the gripper system is heavier [7, 8]. Large-scale production includes also many handling steps, and the amount of energy wasted by each handling operation accumulates. From the economic point of view, a customized solution should be achieved for the handling process.

The gripper design of a customized solution again involves the risk of uncontrolled edges which could collide with the surroundings. Also, if the holding force of the grippers is too low, the layer could fall off during the handling process [2]. When a gripper system is designed with small individual grippers, the arrangement of the grippers on the part is important to avoid the risks in the handling process mentioned above. Unfortunately, the arrangement of the grippers is often performed based on knowledge and experiences acquired by humans [9]. The quality of the solution therefore fluctuates, and the functionality is uncertain until the gripper system has been set up and tested. This paper therefore will present a first approach to designing gripper systems with small individual grippers for handling textiles. The method is based on simulation data to find design solutions.

A handling task which requires such a gripper system design methode ist the production of fiber-reinforced plastic (FRP) parts. Components of FRP are normally built up of different shapes of semi-finished carbon textile [10, 11]. These shapes could be generated by an industrial cutting table on a large scale. To reduce the amount of waste material, different shapes are arranged side by side [12]. Considering this, it is useful to be able to handle different parts with one gripper system. For a mass production it could be assumed that the shape variation of the textile layers is known in advance and that the shapes will not change in the ongoing production. A gripper system should therefore be easy to reconfigure when a new series comes into production. However, for mass production there is no need to change the gripper configuration at short intervals. Besides, when the design of the gripper system is generated, it must be possible to build up any calculated solution. Such features make a reconfigurable gripper interesting for automotive production applications [13].

2　State of the art to design gripper systems

The existing investigations for the design of a gripper system for flats parts could be classified into different categories [9]. Some present a systematic approach for the development process of the gripper system whereby they pay attention to different priorities like flexibility [9], an assembly process [14, 15] or the separation of different plies [16]. These approaches often aim to help to choose between different possible solutions like the gripper technologies and other subcomponents and present the decision-making process in a structured way. These methods create the concept of the gripper but normally didn't help to define concrete geometric dimensions of the gripper system.

Other investigation focuses on the gripper technologies dimensioning. This is achieved by the formulation of an analytic model or diagrams based on experimental results [17–19]. Such investigations only consider one gripper element at the time. However, the handling of flats parts often requires the use of multiple gripper elements.

A few works present approaches for the arrangement of gripper elements on a flat part. For example, the draping of a textile is influenced by the distance of the gripper elements and is therefore an important factor [20]. A other investigation is optimizing the arrangement of the gripper on a metallic sheet by using a finite element (FE) simulation and Powell's algorithm. For the simulation they need a initial gripper arrangement, but did not mentioned how such a arrangment could be generated. [21]

3   Simulation-based approach to gripper arrangement

It is advantageous if a gripper system is designed to handle a set of different shapes as mentioned above. To design such a gripper system a systematic approch is presented in **Fig. 1**. The first five steps generate a gripper configuration separately for each shape. The valid configuration solutions found for each shape are combined in step six to be implemented with the modular gripper system mentioned above.

In a first step, to generate a gripper configuration for a shape, all necessary information is collected. This includes the geometry of all parts as CAD data, the material, the maximum acceleration in the handling process and the maximum allowed deflection of the parts in the handling process defined by the user. In a second step a first estimated gripper configutaion has to befound to allow a finite FE simulation in step three. The results of the simulation will give detailed information about the expected deflection in the handling process and required holding force of the gripper. In step four, the deflection and required holding forces are analysed. For this purpose the simulated deflection is benchmarked with the user input from step one. The the requiered holding force, on the other hand, is compared with a data base which contains experimental data for the maximum holding force a gripper can apply to the material.

Based on this information, a gripper configuration for a shape is valid or not. If the configuration is valid the configuration process will start at step one for the next shape. If the configuration does not satisfy the requirements an adjustment of the gripper configuration is preformed in step five. The measures in step five can distinguish between adding an additional gripper or displacing existing ones to fit the requirements. If a valid configuration for each shape could be found, the different gripper configurations are combined into one solution in step six. This step tries to use the same grippers for handling of the shapes as much as possible to reduce the amount of grippers.

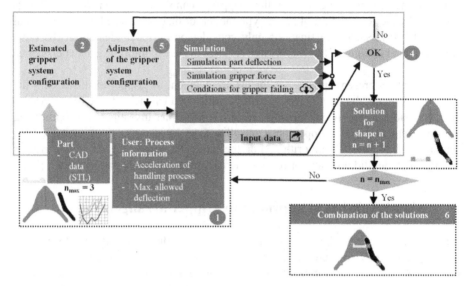

Fig. 1. Systematic approach for the design of the handling system

3.1 Simulation of the textile deflection

The presented approach uses simulated data of the textile deflection to develop a gripper configuration. The deflection is mainly caused by the mass of the textile itself, the acceleration of the handling process and draft. The first step is to simulate the deflection caused only by the mass of the textile itself through the acceleration of gravity. For this purpose, a finite element simulation in Abaqus was set up. The forces interacting with the textile in the handling process are assumed to be low which should allow to use a linear elastic behaviour. In addition, the thickness of a textile is thin compared to the other dimensions of the part, hence the problem should be described by a two-dimensional stress. Because of this, shell elements with a linear elastic material definition called LAMINA are used. The model need two elastic moduli (E_1, E_2), three shear moduli (G_{12}, G_{13}, G_{23}) and the Poisson's ratio (v_{12}). The shear modulus G_{13}, G_{23} should be irrelevant since the material is not very thick. The Poisson's ratio (v_{12}) is set to zero to ignore transverse strain through normal strain. The test material is the plain-weave fabric Sigratex C W160-PL1/1. To measure the elastic modulus a cantilever test according to the standard DIN 53362 is performed. The shear modulus on the other hand is, determined by a test handing process. An equilateral triangle with a side length of 300 mm is gripped and lifted very slowly so that deflection is only caused by the mass of the textile. The deflection is then measured by a laser scanner.

By comparing these measurements with simulation results an experimental shear modulus could be calculated. The exact values obtained from these measurements are listed in **Table 1**.

Table 1. Parameters for Sigratex C W160-PL1/1

Parameter	Value	Unit
E_1, E_2	245,79	N/mm²
G_{12}	0,189	N/mm²

The question is how well such a simulation model with the elaborate parameter values above can fit for any kind of geometric shape. For the latter purpose, a more random shape was created and gripped under the same circumstances as in the case of the equilateral triangle described before. This experiment was repeated six times, and the deflection of the textile was measured with the laser scanner device. **Fig. 2** shows a comparison between simulation and the scans. Overall the difference between simulation and scan remains small. Only in two areas, significant errors occur. Error one (**Fig. 2**) occurs in an area with a very high deflection up to -126 mm in practical handling tests. The simulation predicts a deflection of just -97 mm. Considering the high degree of deformation the error is acceptable and only error two is relevant. At this point the simulation should show -20.24 mm but is calculated to be at -27.55 mm. Nevertheless, the model so far allows to get a goog idea of the textile deflection for a given gripper configuration.

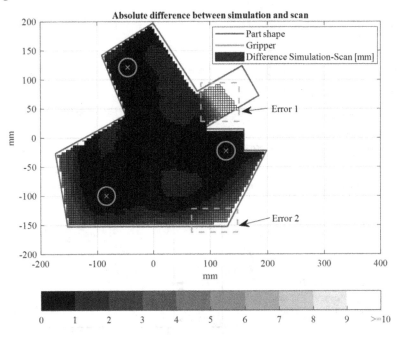

Fig. 2. Absolute difference between simulation and scan

3.2 Method for estimation of a gripper configuration

An important step for the approach presented above (**Fig. 1**) is the estimated gripper configuration shown in step two. Without gripper configuration, a FE simulation is not possible. Furthermore, if the estimated gripper configuration already represents a valid solution or is very close to such a solution the approach becomes effective. The idea is to include simulation data already in the dimensioning process of the estimated gripper configuration. The steps for the estimated gripper configuration are presented in **Fig. 3**.

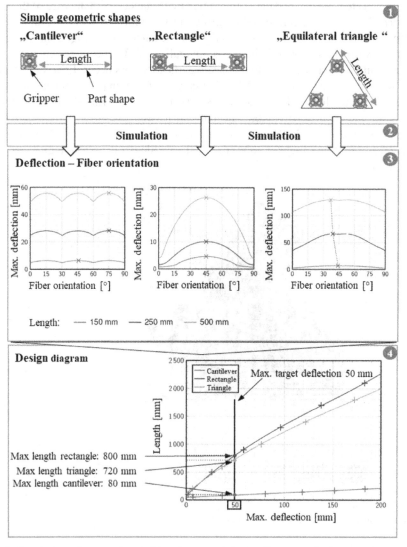

Fig. 3. Simulation of geometric primitive shapes and design diagram to find an estimated gripper configuration

For this purpose, simple geometric shapes are defined e.g., a cantilever beam, a rectangle with a gripper on each end and an equilateral triangle with three grippers. Each geometry is calculated for different dimensions (**Fig. 3** "length") and fiber orientations of the textile. Especially for the cantilever beam and the rectangle, the fiber orientation has a huge impact. Such simulation can run before the shape of a textile is already known and the information could be used to design the gripper configuration for any part with this material. The information gained from the different simulations is combined in one "design diagram" containing the maximum deflection (*DMax*) for each simple geometry and length. From this, the maximum "length" for each simple geometry for a planned deflection (for example 50 mm) could be read.

With the design diagram an estimated gripper configuration is designed for the textile shape shown in **Fig. 4**. The goal is to achieve a *DMax* of 50 mm. The design diagram depicted in **Fig. 3** suggests using triangles with a maximum length of 700 mm, rectangles with a maximum length of 850 mm and cantilever beams with a maximum length of 80 mm. Hence, the goal is to put these simple geometries in the shape of textile while simultaneously obtaining certain maximum dimensions specified by the design diagram. Still, only a minimum number of grippers should be used and, therefore, a minimum number of simple geometries. Mostly, these criteria cannot be fulfilled at the same time or must be solved by a compromise. This is the case for the shape represented in **Fig. 4**. The complete textile could be covered by one triangle and two rectangles. Unfortunately for such a gripper configuration the triangle must be bigger than recommended (790 mm). On the other hand, certain spaces of the shape are not included by the simple geometries. To test this gripper configuration the deflection for different fiber orientations is simulated, and the maximum values are presented in **Fig. 4**. The *DMax* over different fiber orientations remains within given limit of 50 mm between 0° - 15° and 75° – 90°. If the location of the *DMax* is in the area covered by the simple geometries, then the value of the deflection also remains below 50 mm. For fiber orientation between 15° and 75° the location of the *DMax* is always outside of the simple geometries. These areas were not included in the precalculation by the simple geometries. It must be kept in mind that these configurations are only the first step in the development of the final gripper system design. With the help of these simulation results and precise information about the fiber orientation of textile the gripper can now be relocated to areas with a deflection over the maximum value *DMax* in further investigations.

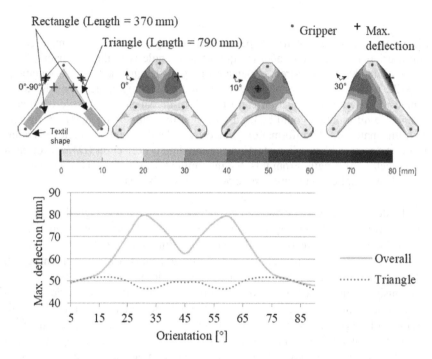

Fig. 4. Simulation of the textile deflection for different fiber orientation for a given gripper configuration

4   Summary and outlook

In this paper, a modular gripper system was presented, and also an approach to develop an estimated gripper configuration was discussed. This approach was to find a gripper arrangement ensuring that the maximum deflection is not exceeded and that only a small number of grippers are used. This approach is based on presimulations of simple geometries which are arranged on the textile shape. This estimated gripper configuration is not perfect for some fiber orientations and exceeds the maximum allowed deflection. Nevertheless, it should be kept in mind that this estimated gripper configuration is only the first step to finding a final gripper configuration in further investigations through iterative optimization.

Acknowledgments

The investigations were part of the research project AsenBa. We wish to extend our special gratitude to the Baden-Württemberg Foundation for founding of the project.

5 References

1. Saadat, M.,Nan, P.: Industrial applications of automatic manipulation of flexible materials. Industrial Robot 29(5): 434–442. doi: 10.1108/01439910210440255 (2002).

2. Fantoni, G.,Santochi, M.,Dini, G. et al.: Grasping devices and methods in automated production processes. CIRP Annals 63(2): 679–701. doi: 10.1016/j.cirp.2014.05.006 (2014).

3. Reinhart, G.,Straßer, G.: Flexible gripping technology for the automated handling of limp technical textiles in composites industry. Prod. Eng. Res. Devel. 5(3): 301–306. doi: 10.1007/s11740-011-0306-1 (2011).

4. Gebauer, I.,Dörsch, C.,Thoben, K.-D. et al.: Automated Assembly of Fibre Preforms for Economical Production of High Performance Composite Parts (2007).

5. Lien, T.K.,Davis, P.G.G.: A novel gripper for limp materials based on lateral Coanda ejectors. CIRP Annals 57(1): 33–36. doi: 10.1016/j.cirp.2008.03.119 (2008).

6. Frederic Förster: Geregeltes Handhabungssystem zum zuverlässigen und energieeffizienten Handling textiler Kohlenstofffaserzuschnitte. Dissertation, KIT Karlsruher Institut for technology (2016).

7. Paryanto,Brossog, M.,Kohl, J. et al.: Energy Consumption and Dynamic Behavior Analysis of a Six-axis Industrial Robot in an Assembly System. Procedia CIRP 23: 131–136. doi: 10.1016/j.procir.2014.10.091 (2014).

8. Rassõlkin, A.,Hõimoja, H.,Teemets, R.: Energy Saving Possibilities in the Industrial Robot IRB 1600 Control: 1 - 3 June 2011, Tallinn, Estonia ; conference proceedings. IEEE, Piscataway, NJ (2011).

9. Straßer, G.: Greiftechnologie für die automatisierte Handhabung von technischen Textilien in der Faserverbundfertigung. Zugl.: München, Techn. Univ., Diss., 2011. Forschungsberichte IWB, vol 256. Utz, München (2012).

10. Angerer, A.,Ehinger, C.,Hoffmann, A. et al.: Automated cutting and handling of carbon fiber fabrics in aerospace industries. In: 2010 IEEE International Conference on Automation Science and Engineering, pp 861–866 (2010).

11. Ehinger, C.,Reinhart, G.: Robot-based automation system for the flexible preforming of single-layer cut-outs in composite industry. Prod. Eng. Res. Devel. 8(5): 559–565. doi: 10.1007/s11740-014-0546-y (2014).

12. Burke, E.,Hellier, R.S.R.,Kendall, G. et al.: Complete and robust no-fit polygon generation for the irregular stock cutting problem. European Journal of Operational Research 179(1): 27–49. doi: 10.1016/j.ejor.2006.03.011 (2007).

13. Tai, K.,El-Sayed, A.-R.,Shahriari, M. et al.: State of the Art Robotic Grippers and Applications. Robotics 5(2): 11. doi: 10.3390/robotics5020011 (2016).

14. Gutsche, C.: Beitrag zur automatisierten Montage technischer Textilien. Zugl.: Berlin, Techn. Univ., Diss., 1992. Produktionstechnik - Berlin, vol 115. Hanser, München (1993).

15. Ehinger, C.: Automatisierte Montage von Faserverbund-Vorformlingen. Zugl.: München, Techn. Univ., Diss., 2012. Forschungsberichte IWB, vol 268. Utz, München (2013).

16. Seliger, G.,Szimmat, F.,Niemeier, J. et al.: Automated Handling of Non-Rigid Parts. CIRP Annals 52(1): 21–24. doi: 10.1016/S0007-8506(07)60521-6 (2003).
17. Stephan, J.: Beitrag zum Greifen von Textilien. Zugl.: Berlin, Techn. Univ., Diss., 2001. Berichte aus dem Produktionstechnischen Zentrum Berlin. IPK, Berlin (2001).
18. Jodin, D.: Untersuchungen zur Handhabung von biegeweichen Flächenzuschnitten aus Leder mit pneumatischen Greifern. Diss., (1992).
19. Böger, T.: Beitrag zur Projektierung von Greifelementen für die Handhabung flächiger, biegeweicher Materialien (1997).
20. Kordi, M.,Husing, M.,Corves, B.: Development of a multifunctional robot end-effector system for automated manufacture of textile preforms. In: 2007 IEEE/ASME international conference on advanced intelligent mechatronics. IEEE, pp 1–6 (2007).
21. Ceglarek, D.,Li, H.,Tang, Y.: Modeling and Optimization of End Effector Layout for Handling Compliant Sheet Metal Parts. J. Manuf. Sci. Eng. 123(3): 473. doi: 10.1115/1.1366682 (2001).

Automated Additive Manufacturing of Concrete Structures without Formwork - Concept for Path Planning

Serhat Ibrahim[1], Alexander Olbrich[1] Hendrik Lindemann[2,a], Roman Gerbers[2,b], Harald Kloft[2,a], Klaus Dröder[2,b], and Annika Raatz[1]

[1] Leibniz Universität Hannover,
Institute of Assembly Technology,
An der Universität 2, 30823 Garbsen
{ibrahim, raatz}@match.uni-hannover.de
[2] Technische Universität Braunschweig
[a] Institute for Structural Design
[b] Institute of Machine Tools and Production Technology

Abstract. At the Digital Building Fabrication Laboratory in Braunschweig, automated additive manufacturing of concrete structures without formwork is researched. The system consists of a six-axis robot and a three-axis milling machine, which are each connected to a three-axis portal. At the robot a shotcrete sprayer is installed to generate concrete structures. In this paper a first path planning concept for the robot is described. The given algorithm calculates from the CAD data of an object a path for the robot. A difficulty of path planning is that the shotcrete application depends on various parameters. Therefore, a simplified model for the shotcrete application is developed. For the purpose of error minimization while spraying, an online monitoring approach by using a laser scanner is presented.

Keywords: additive manufacturing, concrete spraying, path planning

1 Introduction

The concrete industry has made some major developments in the last few years. Ultra High Performance Concrete (UHPC) and fiber reinforced concrete are just some of the high performance materials which are able to create resource efficient building elements [1]. Unfortunately, available casting systems are developed for standard solutions and do not allow the complexity needed to use these materials efficient and therefore cost effective.

The research project *A robotic spray technology for generative manufacturing of complex concrete structures without formwork*, is a cooperation of the Universities of Hannover, Clausthal and Braunschweig and is exploring the possibilities to manufacture freeform concrete structures without a static formwork [2,3]. The core of the project is the Digital Building Fabrication Laboratory (DBFL). The DBFL consists of a six-axis robot and a three-axis milling machine, which

© Springer-Verlag GmbH Deutschland, ein Teil von Springer Nature 2018
T. Schüppstuhl et al. (Hrsg.), *Tagungsband des 3. Kongresses Montage Handhabung Industrieroboter*, https://doi.org/10.1007/978-3-662-56714-2_10

are each connected to a three-axis portal. The robot and the milling machine can be operated either independently or in a cooperative sequence.

So far, the most advanced approaches in 3D printing with concrete comprises extrusion deposition techniques [4]. Spraying the concrete opens up new possibilities to apply the material. Instead of slicing a 3D-object in 2D-layers and stack them up on each other, the additive process can be developed in a real three dimensional manner. In addition to the 3D printing process, the DBFL allows several post processing methods. Already 3D printed elements can be revised by subtracting or adding material using the milling machine or the robot, respectively. First results are shown in [5], where indicate another benefit of concrete spraying. Due to the compacting energy of the spraying process, there seems to be no reduction in strength between the applied concrete layers, which is a major problem in the Contour Crafting process.

For first works, a offline path planning algorithm for the robot is developed. In this case, a method is shown how simplified geometries are made with the DBFL. The shotcrete application is not homogeneously, therefore two concepts for online monitoring are shown.

2 Methodology

2.1 Shotcrete Application Model

In contrast to conventional 3D-Printer or Contour Crafting applications [6–8], the dimension of the applied shotcrete depends on various parameters that are not constant during the process. In general, it depends on the spray distance, the volume flow of the concrete and the path velocity of the robot [9]. Therefore, a model of the shotcrete application is needed for the path planning algorithm. Figure 1(a) shows the parameters of the cross sectional area of a shotcrete layer. This area can be divided into a main layer A and a transition layer T. In doing so, the main layer approximates a rectangle

$$A = hd, \tag{1}$$

where h denotes the height of the layer and d the width. Since the transition layer is very small compared to A, it is neglected. The cross sectional area geometry of the shotcrete application depends on the volume flow $\dot{V}$ of the concrete, the path velocity v of the robot, the nozzle distance a and the spray pressure p. By considering the volume flow $\dot{V}$ and the path velocity v, the area of the main layer in Eq. 1 can be calculated as

$$A := hd = \frac{\dot{V}}{v}. \tag{2}$$

According to *Moser* [9] the layer width depends linearly on the nozzle distance. In this case, the usual nozzle distance is $1\text{ m} < a < 2\text{ m}$. In the given application, the usual nozzle distance will be $0.15\text{ m} < a < 0.3\text{ m}$. At this distance the air pressure has also an influence on the layer wideness, because concrete can be blown away. As a result of this, the pressure becomes an important parameter for the shotcrete application. If the pressure is too low, the compound strength of the concrete will be too low and the strength is negatively affected. If the pressure is too high, then the spread of the concrete will be too high, which impairs the spray accuracy [10].

For modeling the spraying process, experiments with different analyzed settings are performed to analyze the geometry of the resulting layer. The surface of concrete is partly uneven, which exacerbates the measurement of the layer width. For each layer, several spots are measured so that the mean value of the width can be recorded. Figure 1(b) shows the layer width depending on the nozzle distance at different air pressures. From the experiments ,it can be concluded, that the layer wideness increases with the air pressure. Increasing the air pressure from 2 bar to 4 bar results in a difference of the wideness of ≈0.02 m.

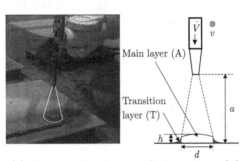
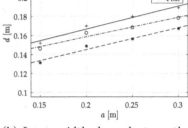

(a) Cross sectional area and parameters of the layer of the shotcrete application

(b) Layer width dependent on the nozzle distance

Fig. 1. Shotcrete application model

Table 1. Interrelation between the nozzle distance and the layer width for different air pressure values

p	b	m	R^2
2	0.099	0.23	0.9718
3	0.118	0.205	0.9588
4	0.118	0.242	0.9708

To find a correlation $d = f(a)$, an interpolation with $f(a) = ma + b$ seems to be a good approach. Here, m and b represent the slope and the y-intersection of the linear function, respectively. Table 1 shows the interrelation between the nozzle distance and the layer width for different air pressure values and the coefficient of determination R^2. The coefficient of determination R^2 of the three experiments are $R^2 > 0.95$, which indicates the practicality of the linear approximation.

2.2 Offline Path Planning

The basic idea of the algorithm, which is presented in this paper, is similar to path planning algorithms used in additive manufacturing [11, 12]. The algorithm reduces a given 3D CAD model to a predefined number of 2D-planes.The major part of the path planning can thus be simplified. After generating the 2D-planes, an individual path is designed so that they can finally be connected. The layer thickness is determined by the required precision [11]. In this case, h_l has to match the optimal application thickness h_{opt}. Optimal thickness is considered to provide the best possible material homogeneity. With an overall height H of the model n planes are created. With h_{max} being the maximal application height, which is technically feasible, and the remainder r of $\frac{H}{h_{opt}}$, n becomes

$$n = \begin{cases} \left\lfloor \frac{H}{h_{opt}} \right\rfloor & \text{,if} \quad r = 0 \quad \text{or} \quad h_{opt} + r < h_{max} \\ \left\lceil \frac{H}{h_{opt}} \right\rceil & \text{,otherwise} \end{cases} . \tag{3}$$

Depending on $h_{opt} + r < h_{max}$, $\frac{H}{h_{opt}}$ is either rounded up or down. If $r \neq 0$, the final layer height becomes

$$h_n = \begin{cases} h_{opt} + r & \text{,if} \quad h_{opt} + r < h_{max} \\ r, & \text{otherwise} \end{cases} . \tag{4}$$

h_n depends on the remainder r and the optimal and maximal application height. The application height can be changed by modifying the velocity of the robot (shown in Eq. (2)).

An ideal path would provide homogeneous concrete application. However, when distortion after application and imperfections, which are caused by the geometry of the path, makes it difficult to achieve. On a straight path on the starting and ending points, less concrete gets applied than in between. When concrete is applied on a curved path more material is applied on the inside than on the out side of the path, which especially takes effect on closed curves, like a circle. In Fig. 2, three different methods, which are provided by the algorithm for computing a 2D-path, are shown. Which method for a plane is used, depends on its width w. If $w \leq 2d_{max}$, where d_{max} is the largest possible layer width of the shotcrete application, an outline path is sufficient to manufacture the entire plane (Case 1). Independent of the used method, the outer path is always calculated by generating a certain amount of points on the planes contour and moving them along a vector. The vector $\mathbf{x} = \hat{e}_z \times \mathbf{t}$ is the cross product of the unit vector $\hat{e}_z$ and the curves tangent $\mathbf{t}$ at the current point, making the computed vector point towards the planes center point. If $3d_{min} \leq w_{min}$ and $w_{max} \leq 2d_{min} + d_{max}$, where w_{min} and w_{max} are the width of the plane at its most narrowest and most widest spot, an additional part of the path finishes the remaining interior area (as illustrated in Case 2). While the outer path is manufactured with a constant minimal application area in order to minimize inaccuracies, the area for the inner is adjusted to the width of the remaining inner plane. The size of the application area is altered by modifying the nozzle distance (Fig. 1(b)).

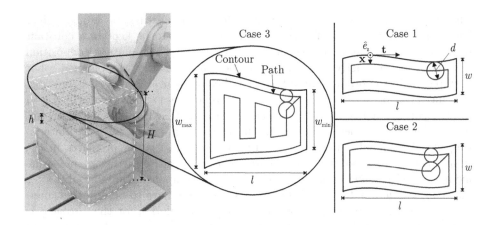

Fig. 2. Path planning cases for the shotcrete application

The points of the path are generated along the inner planes long side and centered by the local width, making the width the diameter being used to apply concrete at this point. In Case 3, $w_{min} \geq 3d_{min}$ applies again and in addition this time $w_{max} > 2d_{min} + d_{max}$ holds. The remaining inner plane is filled with a rectangular path. Again, the outer path is first manufactured with the minimal application area, but the interior is made with the diameter of the application area

$$d = \min\left(\left\{\frac{l}{\left\lceil \frac{l}{d_{opt}} \right\rceil}, \frac{l}{\left\lfloor \frac{l}{w_{min}} \right\rfloor}\right\}\right). \tag{5}$$

d is given as a multiple of d, close to d_{opt} or w_{min}, adds up to the interior planes length l. In order to control the robot in Case 3, only the corner points of the path are required, as the diameter stays constant and the path between the corner points is a straight. The distance between these points along the long side is determined by the calculated diameter, because the alignment to the short side is adjusted to half the diameter to the planes edge. Even though, this method is less precise, but it allows a wider range of objects to be manufactured. As long as the the length l_{max} of the plane exceeds the width w_{max} and the short edges are straight, every object can be manufactured. In particular, it should be mentioned, that the third method allows the manufacturing of objects with a variable width. The transition from the outer to the inner path in Case 2 and Case 3 applies excess material, however, a transition with maximal velocity reduces the effect to a minimum. Each generated point is associated with a height and a diameter, specifying how much concrete needs to be applied in that area.

2.3 Online Path Planning

The application of sprayed concrete for additive manufacturing is a major challenge, as a large number of process parameters have to be kept in a constant ratio in order to achieve an acceptable part accuracy. The material deposition must correspond exactly with the application model so that the planned robot path results in the production of the desired component. However, different disturbances can be observed during the sprayed concrete application, which lead to an inconsistent material deposition. Due to the weight of the material, unpredictable displacements can occur during the application, such as material slippage. Additionally, uneven concrete mixtures, fluctuating pressures in the supply hoses and suboptimal path parameters can lead to further manufacturing inaccuracies. Due to these inaccuracies, many components are unusable after direct production and the path planning parameters have to be adjusted manually in a complex iterative process. One possible solution for this problem is to adapt the planned robot path during the deposition process to compensate complex geometric inaccuracies. Such an online path planning requires live sensor data of the inaccuracies to be compensated. The measuring method should deliver sensor data with acceptable accuracy even in a wide measuring range, has to be robust against different ambient conditions (e.g. dust) and should offer fast data acquisition. Therefore, laser triangulation was chosen for measuring the component geometry in this project since it can better meet the requirements compared to other 3D measuring methods (e.g. Time of Flight).

Two concepts have been developed to incorporate the laser sensor for online path control (see Fig. 3). In Concept 1, a robot path is created offline from a CAD component (see section 2.2). The robot path is then transferred to the motion control system of the DBFL that is run by a Siemens Sinumerik 840D to control the robot and the portal axis. Afterward, the production process is started. The motion control system processes the planned path step by step, while the material is deposited in the planned layers. Thereby, a Stemmer C5 laser triangulation sensor is used to measure the component while the concrete is applied. The sensor data is send to a Beckhoff control system and synchronized with the current robot position to enable measuring in six dimensions. The controller is running a path modification algorithm that takes the planned robot path and the sensor data into account to calculate adjustments for the robots velocity and nozzle distance. The results of this process are send to the motion controller to adapt the robot path online. The data acquisition of the second concept is similar to the first concept. However, the sensor data is used for online generation of a new path. The path planning algorithm, that was priory only used for offline path planning is now used for online path generation during the manufacturing process. Therefore, the algorithm is executed in the real-time environment of the

Beckhoff control using a fixed cycle time. As input of the algorithm a difference model is used in which the measurement data of the laser scanner is merged with the CAD model. In each control cycle, this difference model is.used for path planning of the next robot positions.

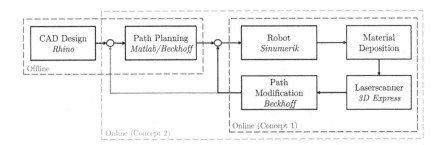

Fig. 3. Concepts for online path planning for increased manufacturing accuracy

The advantage of the first concept is that the necessary algorithm for path modification can be implemented with low effort. However, the concept has the disadvantage that only simple path modifications can be made. For example, the adjustment of the nozzle distance can be used to compensate inaccuracies of the layer width in the small range of approx. 5 cm (see Figure 1 b). Therefore, a compensation of larger geometric deviations, as in the case of a material slippage, is not possible. In order to compensate such defects, the entire path has to be adjusted to rebuild the material before the finishing of the component can be continued. For an industrial manufacturing system, such a compensation mechanism is desirable, but can only be implemented with Concept 2. Therefore, it makes sense to first implement Concept 1 and then upgrade to the second concept later.

3   First Results

For the online path planning the first concept is currently being implemented in the DBFL and preliminary tests have been carried out. Figure 4 shows the robot with the laser scanner and the shotcrete nozzle as well as the corresponding measurement data from the manufacturing process. The path modification algorithm that is run on a Beckhoff control calculates the actual layer width d_{is} as well as the difference between the planned and the actual layer height Δh. Both variables are used to continuously calculate a velocity offset Δv for the robot with

$$\Delta v = \frac{\dot{V}}{\Delta h(d_{\text{planned}} - d_{\text{is}})}. \tag{6}$$

The difference between the planned and the actual layer height corresponds to the nozzle offset Δa, so that $\Delta a = \Delta h$. The velocity offset and the nozzle offset are send with a constant cycle time of 6 ms to the motion controller and are incorporated into the calculation of the robot path. As shown in Fig. 4, the measurement with the laser scanner already leads to good evaluable measurement results. The first tests on the DBFL indicate that the online path planning is a good method to achieve a further improvement in the component accuracy.

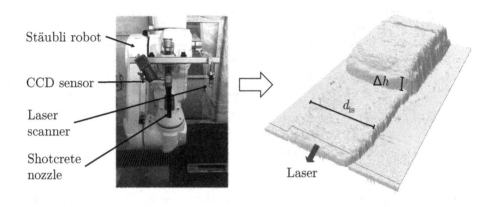

Fig. 4. Current experimental setup for in-process measurement using a laser scanner (left); results of the laser scanner measurements (right)

4 Conclusion and Outlook

In this paper first approaches for an automated path planning for the additive manufacturing of concrete structures were shown. In this case, a shotcrete application model was developed, where the layer width depends on the nozzle distance and the layer height on the velocity of the robot. This offline path planning algorithm uses three different strategies depending on the manufactured geometry. To minimize the application error, two online path planning concepts were described, which use a laser scanner. First results indicate, that the described online path planning improves the accuracy of the process.

For the offline path planning, the effect of the varying robot velocity on the application of the shotcrete should be researched. Especially, if the path has many corners. Here, the online monitoring concept is to be used and examined.

Since additive manufacturing is currently limited by the geometry of the manufactured object, the next step is to expand the application strategy, for example to spray horizontal. An idea is to create a pillar by vertical spraying and then to manufacture the final object by spraying horizontal. In this case, the integration of the laser scanner becomes important to detect application failures during the process, such as material slip off.

References

1. Kusumawardaningsih, Y., 2013: Innovation and Application of Ultra High Performance Concrete. In: Engineering International Conference "UNNES Conversation".
2. Neudecker, S., Bruns, C., Gerbers, R., et al, 2016: A new robotic spray technology for generative manufacturing of complex concrete structures without formwork. In: 14th CIRP Conference on Computer Aided Tolerancing (CAT).
3. Lindemann, H., Fromm, A., Ott, J., Kloft, H., 2017: Digital Prefabrication of freeform concrete elements using shotcrete technology. In: IASS.
4. Hwang, D., Khoshnevis, B., 2015: An Innovative Construction Process-Contour Crafting (CC). In: International Symposium on Automation and Robotics in Construction.
5. Nolte, N., Heidmann-Ruhz, M., Krauss, H.-W-, Budelmann, H., Wolter, A., 2018: Entwicklung von Spritzbetonrezepturen mit steuerbaren Eigenschaften für die robotergestützte additive Fertigung von Betonbauteilen. Accepted for publication in: Spritzbeton - Tagung 2018.
6. Fernandes, G., Feitosa L., 2015: Impact of Contour Crafting on Civil Engineering. In: International Journal of Engineering Research & Technology (IJERT) Vol. 4 Issue 8.
7. Khorramshahi, M. R., Mokhatri, A., 2017: Automatic Construction by Contour Crafting Technology. In: Italian Journal of Science & Engineering Vol. 1, No. 1.
8. Zhang, J., Khoshnevis, B., 2010: Contour Crafting Process Planning and Optimization Part I: Single Nozzle Case. In: Journal of Industrial and Systems Engineering Vol. 4, No. 1.
9. Moser, S.B., 2004: Vollautomatisierung der Spritzbetonanlage- Entwicklung der Applikations-Prozesssteuerung. vdf Hochschulverlag AG, Zürich.
10. Höfler, J., Schlumpf, J., Jahn, M., 2012: Sika Spritzbeton Handbuch 4th edition, Sika Schweiz AG, Zürich.
11. Ding, D., Pan, Z., Cuiuri, D., Li, H., van Duin, S., 2016: Advanced Design for Additive Manufacturing: 3D Slicing and 2D Path Planning. In: InTech.
12. Livesu, M., Ellero, S., Martinez, J., Lefebvre, S., Attene, M., 2017: From 3D Models to 3D Prints: an Overview of the Processing Pipeline. In: eprint arXiv.

Sensor-Based Robot Programming for Automated Manufacturing of High Orthogonal Volume Structures

André Harmel[1,a] and Alexander Zych[1,b]

[1] Fraunhofer Research Institution for Large Structures in Production Engineering IGP, Rostock, Germany
{[a]andre.harmel, [b]alexander.zych}@igp.fraunhofer.de

Abstract. This paper introduces an innovative method for the programming of welding robots in the area of steel volume structure production. The objectives of this programming method are reducing the effort of creating robot programs as well as increasing productivity in production. Starting point of the programming method is a three-dimensional digitalization of the current workpiece. Based on the 3D sensor data of the workpiece the individual components of the construction are identified automatically as well as the weld seams required for connecting them. In this process, 3D sensor data of the components are transformed into simplified regular geometric shapes that will be used later on for collision testing. Collision testing is part of a postprocessor with specially developed path planning algorithms to determine the robot movements required for welding the identified seams. Finally, the robot movements are converted into a system-specific robot program. The developed programming method has been integrated into an existing production facility at Warnow shipyard in Rostock-Warnemünde and was tested under real production conditions.

Keywords: shipbuilding, welding robot, sensor-based programming, collision free path planning.

1   Introduction

Working conditions in the fields of shipbuilding and offshore constructions are characterized by small quantities, large dimensions of components and assemblies and thus by large production and installation tolerances. Under these conditions, automation of working steps is a particular challenge, such as the use of welding robots. Especially robot programming using conventional methods proves itself as inefficient, particularly in the area of one-off production.

Today many shipyards are using welding robots for the production of subassemblies such as orthogonal volume structures (Fig. 1). Respective structures are generally comprised of a bottom plate with stiffeners (panel) and orthogonally arranged girders and floors that are normally also equipped with stiffeners (micro panels). Furthermore brackets, clips and sealing plates are used to connect the individual components. The dimensions of corresponding volume structures such as double bottoms are about 16 x 20 x 3 m, depending on production capacities of the respective shipyard.

© Springer-Verlag GmbH Deutschland, ein Teil von Springer Nature 2018
T. Schüppstuhl et al. (Hrsg.), *Tagungsband des 3. Kongresses Montage*
Handhabung Industrieroboter, https://doi.org/10.1007/978-3-662-56714-2_11

Fig. 1. Assembly of a double bottom section

After manual placing and tack welding the individual components of a volume structure, adjacent components are joined by welding. The vast majority of weld seams can be produced by robots. Finally, remaining seams and, where appropriate, seam ends must be welded manually. In order to program respective robots more efficiently an innovative robot programming method has been developed. This method is based on 3D sensor data of the current workpiece and generates the required robot programs automatically.

2 State of the Art / Related Work

Currently, programming of welding robots in shipbuilding is mainly carried out with offline programming systems (OLP). These systems usually use CAD data of the workpiece to generate the required robot programs [1-2]. However, the available systems have significant disadvantages in practical operation, which hinder the application of robots in many cases:

- Although OLP often has automatic programming features, creating robot programs involves a considerable amount of time and effort, which is usually underestimated by the users. Since a robot program will be used normally only once there is an extremely unfavorable ratio from programming time to program runtime. For this reason, this type of programming has been proven to be inefficient. [3]
- Using OLP requires high quality CAD data. This requirement cannot be fulfilled by all users. In this context, problems occur using different CAD systems for instance when outsourcing construction to external service providers. [4]
- Generation of robot programs by OLP is carried out with a high lead time to the production process. For this reason production becomes inflexible in terms of constructive or technological changes. If such changes occur, robot programs have to be adapted during production process associated with high expenditure of time. Furthermore, offline robot programs have to be adjusted in the production area according to the current position of the workpieces. [5]

Besides conventional offline programming, alternative methods have been developed in the last few years. These methods relocate robot programming back into the area of production and use sensor data of industrial 2D cameras of the current workpiece position and geometry. The major disadvantage of these methods is the fact, that programming also needs either CAD data of the components or manual user input [3, 6-7]. When using CAD data the drawbacks described above come to bear. On the other hand manual user inputs increases the time needed for programming and may possibly lead to errors.

In the area of micro panel production an alternative programming method has been developed based on 3D sensor data, which does neither need CAD data nor user input. This system has been used in production successfully since 2008 [5, 8-9]. However, the current task will go even further, due to the higher complexity of the workpieces and hence the enhanced requirements in terms of sensor data processing. Furthermore, in contrast to micro panels with good accessibility to the welds because of their lower components height, volume structures require collision free path planning of robot movements.

3 System Description

The developed system is characterized by the fact, that robot programming is integrated into the production process and independent of technological preliminary work as well as CAD data. After placing and tack welding of the individual components, the entire volume structure is digitalized using a 3D sensor. Sensor data is then processed automatically to determine the parameters of the seams to be welded as well as a simplified geometric description of the individual components.

This data is then presented to an operator by means of a graphical user interface. The operator is then able to start individual jobs, which will be sent to a postprocessor. The postprocessor determines collision free robot movements for each weld seam as well as switching commands and converts them into a system-specific robot program. Finally, robot programs are transferred to the respective robot controllers and the current welding job is started. Fig. 2 shows the individual components of the modular sensor based robot programming system. The developed system can easily be adapted to different production facilities.

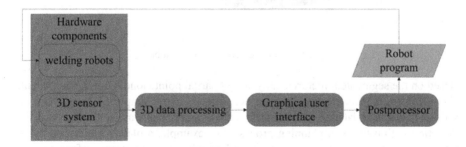

Fig. 2. System components for the sensor-based robot programming

4 Acquisition and Processing of 3D Sensor Data

The most significant characteristic in terms of 3D data acquisition of the workpiece is to minimize shadowing effects to reduce the proportion of manual rework. For that reason, a two-stage sensor data acquisition procedure was developed. In a first step an overview scan of the whole workpiece with low resolution is performed. Based on the data of the overview scan, plates of girders and floors are identified and the workpiece is divided into individual boxes.

In a second step, the individual boxes are digitalized with fine resolution. With knowledge of box positions and dimensions it is possible to determine optimized sensor positions that ensure minimal shadowed areas. Depending on the individual complexity of the current workpiece it is possible to choose different scan strategies.

The methods for processing 3D sensor data have been developed with particular attention to high robustness. This is necessary as the sensor data to be processed are in general inhomogeneous, unorganized point clouds, which are partially affected by erroneous measuring points. Against this background, a multi-stage method for segmentation and classification of 3D point clouds was developed. This method is able to identify the individual components of the scanned box in form of partial point clouds and sort them to generate a hierarchical workpiece structure. This segmentation method is based essentially on an approach to describe components by regular geometric shapes as well as identifying them by means of orthogonal least square distance fitting [10]. Fig. 3 gives an example of a 3D point cloud of a box, which has been segmented into its individual components.

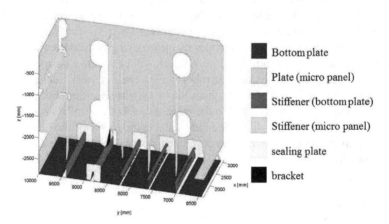

Fig. 3. Segmented components of a box

Based on the segmented, hierarchically sorted partial point clouds of box components, simplified component descriptions are determined. These simplified component descriptions are used later for visualization of box components as well as for collision detection within the path planning process. For example, a plate is described by the plate contour (closed polygon course), bounding cubes and a freeform surface given by knot points to take the deformation of the current plate into account.

Finally, the seams to be welded and their parameters are determined by analyzing the hierarchical component structure and neighborhood relationships. Different weld seams are assigned to different seam types according to the types of components to be joined. By introducing seam types, it is also possible to consider application-specific characteristics such as overlapping seams or splitting weld seams e. g. in order to realize an initial welding of the upper box corners to minimize deformations.

5   Automated Path Planning with Collision Detection

Ensuring collision avoidance by using exclusively CAD data to plan robot movements cannot be guaranteed and, for this reason, needs interaction of the operator. On the one hand, constructive changes e. g. additional components cannot be considered when generating offline robot programs. On the other hand, deviations between CAD data and the actual existing workpiece occur due to production and installation tolerances, for example because of previous welding work. Therefore, path planning methods have been developed that use geometric information of the actual components, obtained from sensor data processing, to determine collision free robot movements for welding. Collision testing is implemented between the simplified component descriptions and the current robot structure. By integrating path planning into the production process, high demands are made with regard to the calculation speed of these path planning algorithms.

The basic approach of collision detection is to perform a distance calculation between two objects in Cartesian space. The most primitive form used here returns a Boolean value that indicates whether there is a collision or not. Calculation of the minimal distance between two considered objects is not performed because of the significantly higher computation time. Further reduction in computing time is achieved by two approaches both applied to the component descriptions and the robot structure:

- reduction in the effort per collision test by the use of primitive approximation
- reduction of the number of collision tests by the use of binary trees

The primitive approximation replaces objects with fully enclosing geometric primitives such as spheres or cubes. Therefore collision detection can be accelerated by using simplified algorithms. Thereby, an approximation error converges depending on the level of detail. This effect is used to reduce the number of intermediate poses to be tested along a robot path by introducing a safety distance on the approximation geometry. To simplify component description cubes are used. The approximation of robot segments is realized by using capsules that reduce the approximation error especially for the welding gun and the robot arms. Moreover, for each robot pose it is necessary to transform robot capsules according to the actual joint angles.

Further reduction of computing time is achieved by using bounding boxes. A bounding box represents a special case of a cube, where the edges are aligned to the axes of the coordinate system, which simplifies distance calculation. For this reason, bounding boxes are used to approximate geometric components on the upper levels of binary trees described below.

Binary trees are hierarchical structures with a root (node) branching out by passing several nodes until the leaves are reached. A node owns a maximum number of two children nodes. The binary tree of the robot structure for example has leaves that correspond to the capsules. The next level consists of bounding boxes for each leave (respectively capsule). By successively merging two bounding boxes each to a new bounding box on the next level a new node is created until the root is reached and the tree data structure is completed (Fig. 4). The tree data structure of the workpiece description is identical except for the leaves, which uses cubes as a primitive approximation.

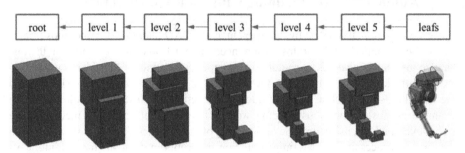

Fig. 4. Binary tree and primitive approximation for a robot structure

Using binary trees results in a reduction of the number of collision test needed. Collision testing starts with the roots of the two binary trees. If a collision between two nodes is detected, four possible combinations of the four children nodes are considered. This procedure is repeated until an absence of collision is detected or the leave level is reached. Only on this level the time consuming test between cube and capsule has to be performed. On the other levels the collision test is performed between two bounding boxes.

The general path planning task deals with the fundamental problems of an over-determined system. The degrees of freedom with regards to a welding gun pose consist of the orientation around the welding wire, the work angle and drag angle of the welding gun as well as the position of the robot base, which can be relocated by means of usually three external linear axes. This results in a large variety of possible robot configurations for each welding gun position along the weld seams. For this reason, an efficient strategy for path planning was developed.

In a first step, path planning is performed only with regard to the welding gun along the current weld seam. To find an appropriate torch pose, a discrete number of positions are analyzed by varying the orientation of the welding gun within the process limits. This process starts with a preferred gun orientation that depends on the current weld seam type.

After finding an appropriate path of the torch along the weld seam the more time consuming process of determining a matching robot configuration starts. In this process a discrete number of robot base positions are analyzed by creating a grid of points in a cylindrical coordinate system. The coordinate system origin is set to the wrist center point of the robot. Grid angle, radius and height are chosen depending on the current weld seam type. In this process of path planning further restrictions occur with regard to:

- A reduction of the hose package twisting
- An avoidance of singularities and changes in robot configuration
- An avoidance of large movements of single robot axis

After finding an appropriate path along the weld seam, the last step of path planning is to calculate the robot paths to reach or to leave weld seams (transition moves). To this end, predefined movement patterns to reach suitable safety positions are used. In some cases, this movement patterns needs to be adapted due to poor accessibility e. g. under horizontal stiffeners in the bottom area (Fig. 5). In this case the robot movement is adapted using different strategies based on the type and position of the component causing the collision along the transition path. Basically, attempts are made to find a path to the current weld seam directly from the last safety position. If this is not successful, an intermediate position based on the current box dimensions is used to reconfigure the robot arms.

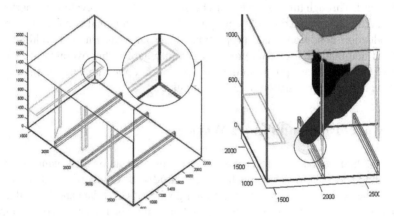

Fig. 5. Left: problematic weld seam / Right: collision at safety position behind the seam end

6 Integration of System Components into a Production Line

The developed system components were integrated into a production facility at Warnow shipyard in Rostock-Warnemünde (Fig. 6). This facility consists of a gantry with two ABB welding robots (IBR 1400), each one having three additional linear axes. Originally the programming of these robots was done based on macro technology using the ARAC software developed by Kranendonk. To implement sensor-based programming, one robot was equipped with a 3D-laser scanner (FARO Focus X130). For 3D data processing and generation of robot programs a computer was provided and connected to an ARAC network drive for communication with the robot system.

Fig. 6. Integration into the production line

The functionality of the developed system was tested under real production conditions using special designed mockup structures reflecting the pluralistic design spectrum of the shipyard. Through this research it has been proven that the developed programming system is capable of realizing robot programming for orthogonal volume structures under production conditions. The proportion of currently determined weld seams by means of automatic sensor data processing with regard to the overall weld length is about 96 %. The average programming time per box is approximately six minutes compared to an average production time of 70 minutes.

7   Conclusion and Future Work

A new method of sensor based robot programming for automated manufacturing of high orthogonal volume structures has been introduced. Compared to state of the art methods the system allows robot programming without CAD data and manual user input. The integration of robot programming in the production process in connection with automatic sensor data processing and collision free path planning based on real workpiece data result in a reduction of programming effort and an increase in production efficiency. The developed programming method has been successfully tested under real production conditions. The currently used robot and system parameters can easily be adapted to other production facilities. Therefore, the developed system is flexible in its use further in terms of designing a new production plant and to upgrade an existing production system.

Future work has the target to extend the areas of application in terms of a wider range of components and weld seam types to be identified as well as to further increase the speed of data processing. Furthermore, it is now possible to update CAD data of manufactured workpieces based on the acquired 3D sensor data and implement quality gates in production. These quality gates ensure that defined workpiece tolerances are respected before the current workpiece is transported to the next production step. This minimizes manual rework on later production steps, which has proven to be extremely time consuming.

Acknowledgments

The project upon which this publication is based was funded by the Federal Minister of Education and Research under the funding code 03 IP T5 09 A. The authors of this article are responsible for its contents.

References

1. Boekholt, R.: Welding mechanization and automation in shipbuilding worldwide. Abington Publishing, Cambridge (1996)
2. Collberg, M., Schmidt, B., Steinberger, J.: Lösung für die Offline-Programmierung von Schweißrobotern im Schiffbau. In: DVS-Berichte Band 176, DVS-Verlag, Düsseldorf (1996)
3. Holamo, O. P., Ruottu, K.: Machine Vision System-Aided Robot Welding of Micropanels. In: DVS-Berichte Band 237, pp. 516-519. DVS-Verlag, Düsseldorf (2005)
4. Ang Jr, M. H., Lin, W., Lim, S.-Y.: A walk-through programmed robot for welding in shipyards. Industrial Robot: An International Journal 26(5), 377-388 (1999)
5. Zych, A.: Automatische Programmierung von Schweißrobotern in der schiffbaulichen Mikropaneelfertigung auf Grundlage von 3D-Sensordaten. Universität Rostock, Rostock (2010)
6. Gilliland, M. T.: Method and apparatus for determining the configuration of a workpiece. United States Patent 5,999,642 (1999)
7. Veikkolainen, M., Säikkö, J.: Welding arrangement and method. European Patent 1 188 510 A2 (2002)
8. Wanner, M., Zych, A., Pfletscher, U.: Method and device for controlling robots for welding workpieces. International Patent WO 2008/101737 A1 (2008)
9. Wanner, M., Zych, A., Pfletscher, U.: Automatic Robot Programming System for Welding Robots in Micro-panel Production. In: Schiffbauforschung 47(1), pp. 39-48 (2008)
10. Ahn, S. J.: Least Squares Orthogonal Distance Fitting of Curves and Surfaces in Space. Springer Verlag, Berlin Heidelberg (2004)

Finite Element Analysis as a Key Functionality for eRobotics to Predict the Interdependencies between Robot Control and Structural Deformation

Dorit Kaufmann[1,a] and Jürgen Roßmann[1,b]

[1] Institute for Man-Machine Interaction, RWTH Aachen University
Ahornstraße 55, 52074 Aachen, Germany
{[a]kaufmann, [b]rossmann}@mmi.rwth-aachen.de

Abstract. The design and usage of a robot requests knowledge from many different disciplines, like mechatronics, materials science, data management etc. Nowadays, computational simulations are an acknowledged method to assure an effective development process of a robotic system. Nevertheless, those simulations are limited to the analysis of single components. This leads to a negligence of the overall picture, which can be fatal, as the failure of a system is often caused by a defective interplay of different components. The concept of eRobotics proposes a framework for an Overall System Simulation, where all occurring interdependencies are explicitly considered. Still missing is an interaction with Finite Element Analysis, which calculates the structural deformation of a component with respect to the actual load case. This work closes the gap and gives a new key functionality to eRobotics, which allows analyzing the impact of structural deformation on robot control and vice versa.

Keywords: Finite Element Analysis (FEA), eRobotics, Rigid Body Dynamics (RBD), Overall System Simulation, 3D Simulation, Virtual Testbed.

1   Introduction

The development of technical systems including robotics becomes more and more complex. The technologized world nowadays has high demands concerning productivity, efficiency, reliability and usability of robots. Concurrently, a low-cost and fast development process needs to be maintained. Thus, computational simulations are essential to understand, test and optimize a robot before even a prototype is produced. Consequently, they are widely used in the fields of research connected to robotics, as mechanics, electrical engineering, materials science, data processing and many more. This interdisciplinary nature will even rise with realizations in the context of Industry 4.0.

Usually, each involved discipline uses its own simulation tool, which is specialized to solve the respective problem. On the one hand, those optimized algorithms guarantee a high accuracy; on the other hand, explicit teamwork between the disciplines becomes very difficult. This means, that there is a sophisticated knowledge about a single component or operation of a robot, while all interdependencies between components and

© Springer-Verlag GmbH Deutschland, ein Teil von Springer Nature 2018
T. Schüppstuhl et al. (Hrsg.), *Tagungsband des 3. Kongresses Montage*
Handhabung Industrieroboter, https://doi.org/10.1007/978-3-662-56714-2_12

operations are neglected. However, the more complex the whole system gets, the more crucial is the understanding of all interactions in the system. Only with a simulation framework considering all aspects, the functionality of the robot can be assured. This approach is followed in the concept of eRobotics [1]. An Overall System Simulation calculates the kinematic movements and the Rigid Body Dynamics (RBD) of the robot, considers the in- and output from controlling and sensors and simulates the impact of the environment. Another crucial aspect is real-time capability, as it allows to do Hardware-in-the-Loop (HiL)-Testing. Finally, these qualities define a Virtual Testbed [1].

Regarding a robot as a mechatronic system, one aspect is still missing in the current overall picture: the structural behavior of components. Finite Element Analysis (FEA) is a sophisticated and recognized method to simulate structures and is widespreadly used with great success. But again, the results remain isolated of the overall system. Nevertheless, the robot movement and the behavior of the underlying structure are highly correlated. When it comes to vibrations or minimal deformations of the base, the Tool Center Point (TCP) is affected even stronger [2]. Thus, structural effects may even cause failure of the whole system. The underlying interdependencies can only be characterized using an active and dynamic interaction between the Overall System Simulation and FEA, which works completely automated and very fast. This paper describes the concept, implementation and several application scenarios of such an interaction and shows that the obtained results are strongly needed in modern robotics.

2 Related Work

There are some attempts to include structural results in modern robotics. Several groups use FEA either completely isolated or based on certain critical situations recorded in experiments or simulations. In [3], a heavy duty industrial robot is analyzed with RBD to get the maximum reaction force in the joints. This value is taken as the input for deformation and stress analysis with FEA. Another common analysis combining dynamics and FEA is the characterization of eigenfrequencies or vibrations of a system, e.g. for Reconfigurable Machine Tools [4] or Flexible Linkages [5]. Those works have in common, that they present approaches specialized to the respective use case.

Another way to enable an interaction with FEA is a classical Co-Simulation. The approaches develop sophisticated mathematical models and algorithms and are therefore rather theoretical [6-7]. Furthermore, a real-time capable Co-Simulation still faces severe problems [8]. Besides the core field of robotics, there are many other application scenarios related to robotics, where the inclusion of structural results is needed, e.g. automotive engineering [9-10].

3 Key Methods and Software

To analyze the impact structural deformations can have on the movements of a robot, FEA was integrated into an eRobotics framework, which already provides powerful Virtual Testbeds. To understand how the developed interaction is working, the underlying key methods and implementations shall be briefly described.

3.1 FEA

FEA calculates how a component reacts due to influences like force, momentum or thermal load. The output is the deformation, stress, strain, temperature etc. that can be found in the component afterwards. Nowadays, FEA is the standard method to calculate the behavior of structures and it is based on sophisticated mathematical models [11-12]. To perform a FEA, mainly *four steps* have to be followed. The first step is *meshing* the component, i.e. discretize it into a huge number of single elements (see **Fig. 1**).

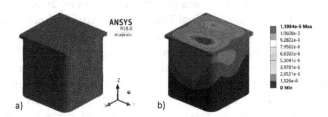

Fig. 1. Two important steps of a FEA, shown for the application scenario of chapter 5: The robot base is meshed (a) and after the solution, the max. deformation in unit [m] can be visualized (b).

For each element, the behavior in the mesh is described by a respectively assigned function. For the overall solution, two conditions must hold: First, the outer forces have to be transferred completely into deformation energy and second, each element i has to keep its connection to the neighboring elements. Thus, a system of m coupled differential equations connecting the displacements u_i and the forces f_i via material parameters k_i has to be solved.

$$\begin{bmatrix} k^1 & & 0 \\ & \ddots & \\ 0 & & k^m \end{bmatrix} \cdot \begin{bmatrix} u^1 \\ \vdots \\ u^m \end{bmatrix} = \begin{bmatrix} f^1 \\ \vdots \\ f^m \end{bmatrix} \equiv k \cdot u = f \tag{1}$$

Solving such a complex system numerically is always a tradeoff between realistic results and an ending calculation time. Thus, the user has to set the parameters for convergence, accuracy etc. manually in the *preprocessing*. The following step of *solving* works completely automated and is very time-consuming: It can last from seconds up to months, depending on the complexity of the model. Once the calculations have finished, the results can be reviewed and analyzed during the *postprocessing* (see **Fig. 1**).

In this work, the FEA part was done with ANSYS Mechanical of ANSYS, Inc. Canonsburg, Pennsylvania. Nevertheless, the general concept of the developed interaction works with any FEA software.

3.2 eRobotics Framework

A robot is a rather complex system as it consists of several components and performs actions in a (maybe non-specified) environment. Consequently, all parts are dependent on each other and have to guarantee not only an isolated, but also an overall system functionality in the interplay with others. Thus, mechanical, electrical, controlling and many more influences have to be considered altogether. An effective way of doing so is using simulations and by that introducing the principle of eRobotics, where complex robotic systems are analyzed virtually.

To assure an extensive and sophisticated Overall System Simulation, a strong framework with many already included functionalities is needed. In [13] such a framework is presented. It is based on a microkernel architecture, called Versatile Simulation Database (VSD), which handles vital functions as data-management, communications and models. Further features of modern robotics are included by extensions, like sensor simulation, kinematics, dynamics, path-planning etc. (see **Fig. 2**). As the environment of the robot is modeled as well, a Virtual Testbed is created, which allows doing design, development, testing and optimization processes in the eRobotics framework. Finally, the simulation is rendered in 3D and comes with a Graphical User Interface (GUI). The underlying algorithms are C++ based and therefore in general real-time capable.

For the developed interaction with FEA, the input forces are needed, which the robot movements implicitly generate. These are calculated by RBD, where a collision hull, the center of mass and the inertia tensor is assigned to each component. The equations of motion of a rigid body (RB) become more difficult, when it is connected to other RB via joints. This defines a forbidden direction of movement, which can be also seen as a constraint force $Z = \lambda \cdot \nabla f$ acting on the component, where ∇f points in the direction of the force and λ is a Lagrangian multiplier. Thus, only velocities $\dot{r}$ perpendicular to ∇f are allowed, which leads to a so-called holonomic constraint with any constant b:

$$\nabla f \cdot \dot{r} = b \iff \nabla f \cdot \dot{r} - b = 0 \qquad (2)$$

Regarding the physical system, the constraint force has to be added to the Newtonian Axiom and one finally gets the Lagrange equation.

$$m\ddot{r} = F_{ext} + \lambda(t) \cdot \nabla f(r, t) \qquad (3)$$

Although the RBD problem is fully described with the equations above, Stewart and Trinkle [14] proposed a momentum/velocity-based approach because it is favored in terms of performance. $\dot{r}$ and $\ddot{r}$ are put in relation to each other with an easy integration $\ddot{r} = \frac{\dot{r}(t+h) - \dot{r}(t)}{h}$ of time step h. Rearranging the terms results in a Linear Complementarity Problem (LCP), which can be solved by various algorithms [15].

$$\begin{pmatrix} m & -\nabla f(r, t) \\ \nabla f(r, t) & 0 \end{pmatrix} \cdot \begin{pmatrix} \dot{r}(t+h) \\ \lambda(t) \cdot h \end{pmatrix} - \begin{pmatrix} F_{ext} \cdot h + m \cdot \dot{r}(t) \\ b \end{pmatrix} = \begin{pmatrix} 0 \\ 0 \end{pmatrix} \qquad (4)$$

This equation finally delivers the constraint forces and momentums Z acting on a RB which are needed for the FEA.

3.3 Concept of the Interaction

The integration of FEA into the eRobotics framework is realized with a bidirectional exchange of characteristic variables (see **Fig. 2**). The constraint forces/momentums Z of any RB are calculated in the eRobotics framework and serve as an input to the FEA.

There, the respective displacements u are calculated and then given back to the eRobotics framework.

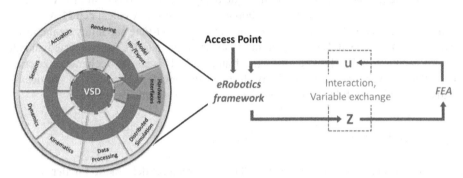

Fig. 2. The eRobotics framework is highly versatile and includes already many features of modern robotics (left, [13]). It is the starting point for an interaction with FEA, which completes the framework. The concept is based on a bidirectional exchange of characteristic variables (right).

4 Operating Mode of the Completed Framework

The novelty of the developed approach is that structural results can be directly considered in the eRobotics framework (see **Fig. 3**). After a FEA was performed and a structural deformation was detected, all RB of the overall system affected by this displacement can be updated to a new pose (as described in [16]) and the deformation can be visualized. Thus, not an isolated and belated static view is given, but the interdependencies between robot movement and structural behavior are dynamically considered.

Fig. 3. Schematic view of the functionality of the developed interface: when a robot performs a task (a, path in yellow), the structural deformation of the base is calculated in a FEA (b, color coding for deformation. Robot manually shifted for visualization).

As it is always crucial for interactions to retain the accuracy, a separation of competences is guaranteed in the interaction process. This means, that the FEA can be setup beforehand by an expert and the user of the eRobotics framework just starts a new calculation in the given setup with the current forces. This is done completely automated after the user presses a button in the GUI. The implementation and dataflow in the background is described in [16], where a validation for retaining the accuracy during the interaction was performed as well. Furthermore, there are several operation modes of the interaction: an overnight mode delays the time-consuming FEA calculations to a suitable time and an interpolation mode uses formerly calculated FEA results. This brings the completed framework much closer to real-time capability again, as no explicit FEA is needed. Besides, maximum loads during a task can be detected and all results can be visualized in the eRobotics framework. This allows to characterize critical situations very fast and intuitively. In total, the developed interaction provides a wide range of powerful analysis tools that are at the same time easy to use.

5   Applications

Every robot interacts with components showing structural deformation under loading conditions. Usually, the displacements are rather small. Nevertheless, through lever action or sudden unexpected forces, they can exceed the permitted dimension and thereby destruct the functionality. Thus, the developed interaction provides a tool to determine which interdependencies robot movement and structural deformation show: *In which cases they are negligible? Which design improvements shall be made?* Three different applications are presented in a feasibility study of a robot performing a simple gripping task. The structural deformation of the base is calculated and the mutual reactions between robot and base are analyzed (see **Fig. 3**).

5.1   Robot Control

Especially robot control has a high impact on the underlying structures, as a badly controlled robot does not move smoothly and therefore creates high peaks in the contact forces of the robot flange bearings, where it is fixed (see **Fig. 4**, upper plots).

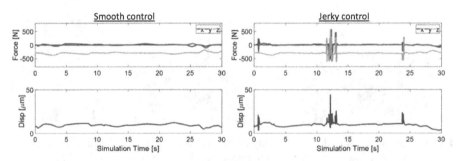

Fig. 4. The influence of robot control quality onto structural deformation: for smooth control, the absolute displacement of the base does not vary much (left), whereas for jerky control, it shows peaks which are more than 3.5 times as high as usually (right).

As those peaks exceed the usual contact forces occurring in a smooth testing, they might not be noticed when designing the robotic system and lead to failure later on. This can be clearly seen in the structural deformation of the base for a jerky control case, which is up to 350 % higher than for smooth control (see **Fig. 4**, lower plots).

5.2 Material Study

Another important question for constructing the robotic system is the choice of materials. The behavior under loading conditions can be easily compared for different materials by integrating FEA results into to the eRobotics framework. The base was once build out of steel and once out of aluminum for a comparison. In **Fig. 5**, the deformation is shown for an exemplary load case occurring during the gripping task of the robot.

Fig. 5. Comparison of different influences on maximum structural deformation in unit [m]. The robot base is slightly varied in the material/design: a) original base made of steel, b) base made of aluminum, c) base made of steel, but with a 50 % thinner upper plate.

5.3 Design Optimization

Apart from the material, design choices have to be made for the construction of a robot. This is especially true, when there are outer limits to consider, like weight, narrow spaces or very expensive materials. As an example, the robot base used in the example was reconstructed virtually with an upper plate half as thin as before. The impact on the structural deformation for the chosen load case can clearly be seen (see **Fig. 5**).

6 Conclusion

In this work, an existing Overall System Simulation for robotics was extended by an interface to FEA. Thus, the occurring structural deformations caused by a moving robot can be dynamically considered. It could be shown, that the controlling of a robot has huge impacts on the loads that act on the underlying structure and therefore may result in non-expected deformation, which might cause failure of the whole system. Similar, the development of bearing or base components can be optimized with respect to material and design, as the loads defined by the robot are obtained by the interface. The developed interaction is a sophisticated approach that adds a new key functionality to eRobotics, since the interdependencies between robot control and structural deformation can be predicted explicitly. Moreover, the model independent design makes the interaction usable for many other applications in the future, as e.g. soft robotics.

Acknowledgements. This work is part of the project "INVIRTES", supported by the German Aerospace Center (DLR) with funds of the German Federal Ministry of Economics and Technology (BMWi), support code 50 RA 1306.

References

1. Rossmann, J., Schluse, M., Rast, M., Atorf, L.: eRobotics combining electronic media and simulation technology to develop (not only) robotics applications. In: E-Systems for the 21st Century – Concept, Development, and Applications (Kadry, S. and El Hami, A., eds.), vol. 2, ch. 10, Apple Academic Press (2016). ISBN: 978-1-77188-255-2.
2. Montavon, R., Zirn, O.: Couplage des modèles d'analyses par élément finis et corps-rigides en machines-outils. (via ResearchGate), (2008).
3. Chung, G-J., Kim, D-H.: Structural Analysis of 600Kgf Heavy Duty Handling Robot. In: 2010 IEEE Conference on Robotics, Automation and Mechatronics, pp. 40-44 (2010).
4. Kono, D., Lorenzer, T., Weiker, S., Wegener, K.: Comparison of Rigid Body Mechanics and Finite Element Method for Machine Tool Evaluation. Eidgenössische Technische Hochschule Zürich, Institut für Werkzeugmaschinen und Fertigung (2010).
5. Wang, X., Mills, J.K.: A FEM Model for Active Vibration Control of Flexible Linkages. In: Proceedings IEEE Int. Conf. on Robotics & Automation, pp. 4308-13. New Orleans (2004).
6. Busch, M.: Zur Effizienten Kopplung von Simulationsprogrammen. Dissertation in mechanical engineering at the University Kassel, kassel university press GmbH, Kassel (2012).
7. Schmoll, R: Co-Simulation und Solverkopplung – Analyse komplexer multiphysikalischer Systeme. Dissertation in mechanical engineering at the University Kassel, kassel university press GmbH, Kassel (2015).
8. Stettinger, G., Benedikt, M., Thek, N., Zehetner, J.: On the difficulties of real-time co-simulation. In: V Int. Conference on Computational Methods for Coupled Problems in Science and Engineering, (S. Idelsohn, M. Papadrakakis, B.Schrefler, eds.), pp. 989-999 (2013).
9. Ambrosio, J., Rauter, F., Pombo, J., Pereira, M.: Co-simulation procedure for the finite element and flexible multibody dynamic analysis. In: Proceedings of the 11th Pan-American Congress of Applied Mechanics (2009).
10. Dietz, S., Hippmann, G., Schupp, G.: Interaction of Vehicles and Flexible Tracks by Co-Simulation of Multibody Vehicle Systems and Finite Element Track Models. In: The Dynamics of Vehicles on Roads and Tracks, 37, pp. 372-384, 17th IAVSD, Denmark (2001).
11. Bathe, K-J.: Finite Element Procedures. Prentice-Hall, Inc., Upper Saddle River, US (1996).
12. Rieg, F., Steinhilper, R. (edts): *Handbuch Konstruktion*. Carl Hanser Verlag München, Wien, pp. 849-857, (2012). ISBN: 978-3-446-43000-6.
13. Roßmann, J., Schluse, M., Schlette, C., Waspe, R.: A New Approach to 3D Simulation Technology as Enabling Technology for eRobotics. In: Van Impe, Jan F.M and Logist, Filip (Eds.): 1st International Simulation Tools Conference & Expo, SIMEX (2013).
14. Stewart, D., Trinkle, J.C.: An Implicit Time-Stepping Scheme for Rigid Body Dynamics with Coulomb Friction. In: ICRA (2000).
15. Jung, T. J.: Methoden der Mehrkörperdynamiksimulation als Grundlage realitätsnaher Virtueller Welten. Dissertation at RWTH Aachen University, Departement for Electrical Engineering and Information Technology (2011).
16. Kaufmann, D., Rast, M., Roßmann, J.: Implementing a New Approach for Bidirectional Interaction between a Real-Time Capable Overall System Simulation and Structural Simulations. In: Proceedings of the 7th Int. Conference on Simulation and Modeling Methodologies, Technologies & Applications, pp. 114-125, Madrid, ES (2017). ISBN: 978-989-758-265-3

Semantically enriched spatial modelling of industrial indoor environments enabling location-based services

Arne Wendt[1,a][0000-0002-5782-3468], Michael Brand[1,b]
and Thorsten Schüppstuhl[1,c]

[1] Institute of Aircraft Production Technology, Hamburg University of Technology
Denickestraße 17, 21073 Hamburg, Germany
{[a]arne.wendt, [b]michael.brand, [c]schueppstuhl}@tuhh.de

Abstract. This paper presents a concept for a software system called RAIL representing industrial indoor environments in a dynamic spatial model, aimed at easing development and provision of location-based services. RAIL integrates data from different sensor modalities and additional contextual information through a unified interface. Approaches to environmental modelling from other domains are reviewed and analyzed for their suitability regarding the requirements for our target domains; intralogistics and production. Subsequently a novel way of modelling data representing indoor space, and an architecture for the software system are proposed.

Keywords: Location-based services, indoor environmental modelling, semantic spatial modelling, intralogistics, production.

1   Introduction

The location-based services (LBS) available on GPS-enabled mobile phones provide value to users by assisting them based on their location and other contextual information. Likewise, one can imagine LBS could prove beneficial to workers and mobile robots in industrial indoor environments as well. Enabling utilization of the wide array of (positional) sensors already present in autonomous systems used today, holds a lot of potential for providing these LBS.

In this paper a dynamic environmental model, named RAIL, is proposed. The model is to be implemented as a software system, aiming to standardize the dynamic representation of industrial indoor environments in an effort to enable a range of LBS.

RAIL features the following constitutive properties:

- Object-oriented environmental model
- Spatial information is represented by relative 6D transformations between objects
- Integration of information from different sensor modalities in a common model
- Implementation of the dynamic map as a software system providing unified interfaces each for feeding the map with data and querying data from the map
- Conceptually supporting any kind of data to enable near arbitrary LBS

© Springer-Verlag GmbH Deutschland, ein Teil von Springer Nature 2018
T. Schüppstuhl et al. (Hrsg.), *Tagungsband des 3. Kongresses Montage
Handhabung Industrieroboter*, https://doi.org/10.1007/978-3-662-56714-2_13

2   State of the Art and Related Work

With the goal of providing LBS for industrial indoor environments, it seems reasonable to look at what is established in the outdoor world today. Outdoor LBS are ubiquitous in today's society, accessible through commercial products, with Google Maps being at the forefront here. Outdoor LBS can serve as an example to get a grasp of the possibilities of LBS in general: Basic outdoor LBS include navigation services, services providing geospatial range querys paired with additionally available data (POIs of a certain type in range, *e.g. "Find open supermarket within 1000 m"*) and user notifications based on geofencing (advertising [1], disaster management [2, 3, 4]). For a broader classification, Barnes [5] divides outdoor LBS into four domains: Safety, navigation and tracking, transactions and information. A similar classification is employed in [1]. It comes out clearly that there is a spectrum ranging from very basic LBS, merely passing on location information, up to LBS extensively combining spatial with other information. Like with outdoor LBS, development of indoor LBS and the assessment of their potential focus mainly on the consumer market. Most commercially available systems target the services navigation and geofencing for user notification (mainly advertising). Examples of commercially available solutions are: infsoft [6], IndoorLBS [7], IndoorAtlas [8], estimote [9], MazeMap [10] and micello [11]; which seem to be predominantly employed in shopping malls. Other applications include disaster management, human trajectory identification and facility management [12]. Taking a look at LBS targeted at industrial applications there are commercially available systems providing the services positioning and identification as well as geofencing for indoor production environments. Two examples of this are Ubisense SmartFactory [13] and Quantitec IntraNav [14]. Both solutions are extensible via their SDK, but remain proprietary solutions.

With the aim of providing LBS, an underlying spatial model of the indoor environment suited for the targeted service-application is needed. Modelling of indoor environments is an active field of research where two predominant use-cases can be described [12]: First, providing context sensitive information and aid to people acting in indoor space. Second, gathering information about the building itself for maintenance applications. In [15] two main types of modelling indoor space are inferred: Namely, geometric and symbolic spatial models. Geometric models are further divided into cell-based and boundary-based representations. With cell-based models, the physical space is dissected into cells, where every cell gets labeled according to its function (e.g. free space, walls, doors). The cell size is chosen to fit the needed accuracy of the application. In boundary-based representations, only the boundaries of a space (e.g. walls, obstacles) are modelled with geometrical primitives. This yields a very compact map, but no semantic information is included. Symbolic models focus on representing the environment and objects of interest in the form of topological-based structures, graphs or hierarchies. That way, relations such as connectivity, adjacency, and containment can be modelled, thus representing the logical structure of a building's components. Object-oriented symbolic models put a special focus on representing individual objects including semantics. This stands in contrast to only representing building components (i.e. rooms, floors). Symbolic models do not necessarily contain any geometric information.

Reviewing literature on the topic of modelling indoor environments we identify the focus as human navigation and assistance in static environments.

Another field that has to be considered when surveying environmental (especially spatial) representations is that of mobile robotics. The maps used by mobile robots focus on enabling specifically localization and navigation. However, they usually don't feature any semantic information nor notion of objects. As well in the field of robotics, the ROS-tf-library [16] is mentionable. Though not necessarily targeted at representing indoor space, it aims at representing (tempo-) spatial relationships in a symbolic form as a tree of relative transformations between coordinate frames. The tf-library does not support modelling anything besides transformations, thus not lending itself to model any environment as a whole by itself – lacking capabilities of modelling geometry and semantics.

3   Our approach

The main goal of RAIL is to simplify development and provision of LBS in industrial indoor environments by providing open infrastructure to be able to model, acquire and share the needed information and data. The open nature of the system is a key element in setting it apart from commercially available solutions, as we believe that removing commercial interest on the platform side will foster the emergence of a more innovative and diverse range of services and the integration of a more diverse range of data sources. For that purpose, in the following sections, a suitable environmental modelling approach, and a concept for the software system implementing this model are proposed. It has to be stressed that RAIL does not describe a data format for storing a persistent spatial model suitable for data exchange - e.g. CityGML [18], GeoJSON [19] - but is a (distributed) software system managing the data representing the current environmental state. This section starts with our take on LBS and the target scope of RAIL regarding LBS-consumers, environment and use-cases. This is followed by an evaluation of existing modelling approaches (outlined in section 2) in terms of the applications requirements. The successive subsections will cover the data modelling and software system's design.

LBS shall provide actors with required information, enabling them to fulfill their assigned tasks. These actors are explicitly not restricted to be human, but include autonomous (robotic) systems as well. Focusing on industrial applications, especially production, intralogistics and MRO (Maintenance, Repair and Operations), the tasks to fulfill will likely require interaction with and manipulation of objects in the environment – assisting interaction with the physical environment being one of the problems in modelling, as identified in [17]. The environment to be acted in is assumed to be a factory floor, including machinery, storage racks, assembly areas, etc. Combining the assumption that tasks involve object interaction/manipulation with the integration of autonomous systems, the need for not only positional information, but for full 6D-poses can be deduced - which in turn becomes a key requirement for the proposed model. To represent factory floors (some hundred meters across) and at the same time enable handling operations in production (which requires (sub-) millimeter resolution), simultaneously a high resolution and wide spatial range are required to be representable by the model.

Checking these requirements against geometrical models, it becomes evident that 6D-transformations of objects can't be stored in the described geometrical models which renders them unsuitable. On top of that, the conflicting requirements of a wide spatial range and fine resolution would likely result in a poor performance with a purely geometrical representation. When checking the requirements against symbolic models, an object-oriented approach seems viable for representing an industrial environment and the individual objects of interest. Geometric information can be included in the form of 6D-transformations connected with each object in the model. Different from the classification in [15], we see no use for a hierarchical structure of an object-oriented model for factory-floor modelling. As opposed to modelling e.g. office buildings, factory-floors usually lack structural elements like floors and rooms, thus not benefiting from hierarchical modelling on basis of containment.

Looking at spatial modelling techniques in mobile robotics, the environmental models usually have no notion of individual objects and lack methods to model semantic information, both of which are key for enabling advanced LBS that aid context-sensitive interaction with objects. Thus, these modelling approaches are unsuitable in the present case.

The requirements to be satisfied by RAIL and the chosen modelling, arising from the targeted applications and industrial environment shall be as follows:

- Provide a dynamic model of the indoor environment
- Represent relative spatial relationships in the form of 6D-transformations
- Integrate (sensory) data from different domains
- Allow deposition of CAD-models, images, feature descriptors, ...
- Robust against shutdown/failure of software components
- Robust against hardware shutdown/failure and partial infrastructure outage
- Consider scalability for data providers, data consumers and amount of data in the model

3.1   Data Modelling

A major problem in designing the data modelling is the demand to enable a as broad as possible range of LBS. The challenge here is that the concrete LBS to be supported are unknown at this point. Since all modelled data serves the goal of enabling LBS, it is unknown what exactly has to be modelled. Thus we will limit RAIL here to only operate on spatial relationships and geometric primitives to enable basic operations such as *spatial-range-queries*, but won't otherwise interpret the supplied data - which shall rather be done by the LBS or their consumers.

With the goal of assisting in interaction with and manipulation of objects in the environment by providing LBS, there is a strong focus on the actual objects making up the environment. We need to be able to either identify these objects or at least categorize them to be of any use for the intended application, as only location and identity of objects combined allow reasoning about the environment on a semantical level. Hence, we propose an object-oriented modelling approach, explicitly modelling all necessary individual objects with their specific properties, thus adding value to the LBS present.

In this context an object shall be an abstract concept. That is, an object can be purely logical and does not necessarily need a physical representation. Object shall be a logical clustering entity comprised of arbitrary attributes and assigned values. This approach lacks a higher level (hierarchical) ordering principle, but by allowing arbitrary attributes to be assigned to objects, we preserve the option to model logical relationships (containment, adjacency, connectivity, etc.) between objects. Conceptually, the location of an object shall be an attribute of the object.

Regarding the implementation of a dynamic object-oriented spatial model, we see the need to separate the spatial information from the remaining attributes of the modelled entities. The first reason for this is the requirement to model relative spatial relationships. Modelling relative spatial relationships can help in reducing uncertainty and increasing accuracy of transformations retrieved from RAIL by lowering the number of individual transformations (each with their associated uncertainty and resolution) required to compute the actually desired transformation. As noted earlier, the spatial relationships/transformations between objects shall all be in 6D, each carrying additional information about uncertainty and resolution. These relationships effectively construct a graph. The nodes of this graph are the coordinate frames of entities other entities' positions have been determined in, thus representing the modelled objects. The edges represent the transformations between these frames. For computations on the graph, the weight assigned to the edges shall be the uncertainty and resolution of the transformations, allowing to retrieve the transformation with the highest resolution and lowest uncertainty between two frames/objects. As we can see here, querying for transformations (i.e. positional data) requires computation apart from that required for performing database searches. This stands in contrast to queries regarding all other information present about modelled entities. This of course bases upon the fact, that we do not directly interpret any other than spatial information, but rather only relay it. The second reason is that we want to enable spatial range queries which depend on the ability to calculate transformations while taking into account the dimensions of entities as well.

In summary, RAIL will consist of a type of document store, holding arbitrary information about the modelled entities, and a graph of all known transformations between these entities. The document store will be holding the entities as containers for the semantics associated with the elements making up the object-oriented model. The graph containing the transformations between coordinate frames of known entities adds the spatial component to the model. All spatial data is held in a symbolic form, thus not limiting accuracy[1] by design.

---

[1] Except for limits posed by computer architectures.

3.2 Systems Design

This section will begin with a presentation of the overall structure and the interfaces to RAIL. This is followed by an overview of the proposed software system's design, targeted at satisfying the requirements of building a dynamic environmental model for industrial applications.

As noted at the beginning of the last section, RAIL has to be tailored in a way enabling it to serve as the data basis for a diverse range of LBS, with the concrete LBS to be supported being unknown at this point in time. The requirement arising from this is that we need to be able to feed arbitrary data into RAIL, as well as be able to retrieve all data present. This requirement allows to reduce the interfaces to two general interfaces; one for data providers and one for consumers. This greatly reduces the complexity of the interfaces, as they both serve a distinct purpose.

With RAIL not only targeted at modelling industrial environments, but to be used in industrial applications, we have to satisfy the demand for high failure tolerance and robustness by design. Further we will try to enable horizontal scaling regarding most parameters influencing performance to cope with increased data volumes and computational load. We will require all interaction with RAIL to be on TCP/IP and UDP, and explicitly do not target operation on a field-bus-level. For the conceptual design of the interfaces we will further make the following assumptions: Data providers are likely to be on wireless networks with low reliability, and consumers requiring high frequency updates, using the publish/subscribe-interface, are on highly reliable connections.

To account for these characteristics of the data providers/sensors, we want the according interface to be state- and session-less on the application-layer to allow connection drop outs without requiring any time consuming re-establishment of a session. Simultaneously enabling flexible data routing and eased hand- and failover of software components. For data consumers, the assumed high reliability connection allows to spare efforts on making the interface state- or session-less. As we are developing a centralized model, there will be no explicit many-to-many messaging from a data providers point of view, meaning we do not need publish/subscribe-messaging on the systems input and data providers shall "push" their data. Therefore, and with the intention of building state- and session-less interfaces, RAIL will not allow data providers to stream their data, but limit all communication on the ingoing side to message-based communications. On the consumer side we will provide a request/response-interface for queries to the model as well as a publish/subscribe-like-interface. The interface provides a "change-feed" to the consumer continuously providing it with updates to the query it posed initially.

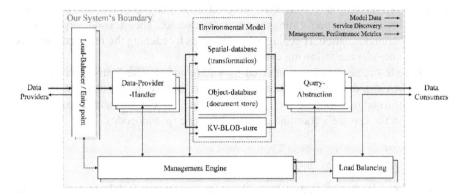

Fig. 1. Overview of the proposed software system. Modules in grey are slaves of the respective module, mirroring its state, for failover.

Targeting the systems robustness and to minimize the effect of crashed instances of software modules, all communication sessions held over the system's boundary shall be executed in their own process – thus not taking down any other instances of the same modules in the event of a crash, opposed to simple parallelizing by threading. All modules shall further be able to be run on separate machines, requiring not more than a network connection, thus enabling horizontal scaling in terms of computation power, memory and network link capacity, and allowing failover to different hardware. For modules only necessary once, in this case the "Load-Balancer/Entry point" (see **Fig. 1**), the "Management Engine" and the consumer side "Load Balancer", there shall be two instances running in master/slave-mode, effectively mirroring each other, on different machines. In the event of an instance failing or connection dropping, the second instance will take over operation, thus introducing a level of robustness and failure tolerance against network and machine failure.

The overall composition of the proposed software system can be found in **Fig. 1**. We will elaborate on the single modules in order of data-flow, from left to right.

Load Balancer, Entry Point.
The load balancer is the entry point from the outside to the system. It will request a new "Data Provider Handler" - on the machine currently assigned to spawn new instances, depending on the available machines loads - when a new data provider supplies data. It keeps track of the assignment of data providers to the "Data Provider Handlers" and routes incoming data to the appropriate handler. This internal routing allows the elimination of session handling in the data providers and provide a session- and state-less interface to the outside world, while still enabling internal caching and load balancing. The load-balancer in its function as entry-point to the system shall provide its address via broadcasts to ease data-provider hook-up, and allow system reconfiguration and migration without explicitly distributing changed addresses.

Data Provider Handler, Lookup caching.
The "Data Provider Handlers" in their simplest form relay the incoming data to the appropriate databases by performing the necessary queries. Their second purpose is ID-lookup and -transformation, speeding up this process by caching the required data. For illustration consider the following example:

A QR-code-marker has been detected by a 2D-camera. The "sensor-driver" (data provider) provides the following information about itself to the handler: ID and type, as well as information about the detected marker: type, ID (content) and the pose (transformation) in the camera's coordinate frame. Message (pseudocode):

```
{ [sensorID:foo, sensorType:camera],
  [type:marker.QR, markerID:bar, TF:<Mat>]   }
```

The problem is, that the sensor-driver does not provide the models internal ID of the marker, this has to be solved with an ID-lookup. The handler's function is to cache the internal IDs of all elements to be recognized by the sensor from the database, to allow fast ID-lookup and data relaying to the spatial database without expensive database queries. This keeps network traffic and computational load down, while also relieving the objects database server – at the cost of increased memory usage on the machines running the data provider handlers. A separate handler is instantiated for each data provider.

Environmental Model.
The spatial model will consist of three databases. One "spatial-database" building the graph of transformations like outlined in section 4.2. One document-store, the "Objects-database", enabling storage of (abstract) objects and assigning arbitrary attributes and corresponding values to them. Last, a separate key/value-store for persistent large binary objects (BLOB) like CAD-models, large images, etc. to take load from the objects data base itself and speed up querying.
The spatial database will ultimately be a custom design, for the remaining two existing technology may be used. Regarding the design of the database we will not elaborat in detail at this point. To satisfy the requirements regarding robustness and failure tolerance, the following requirements are posed on all databases: The databases shall support replication to different machines, automatic failover to different machines carrying copies of the database, and load balancing over their replicated instances or by database sharding. Aiming to support automatic and continuous change-feeds to consumers, we also require the objects- and spatial-database to support change-feeds natively. This eliminates the need for constant polling and querying of the databases with high frequency to supply up to date change-feeds, thus reducing the overall load on the system.

Query-Abstraction.
Query-Abstraction modules are instantiated per individual query to the system. They either process a single query for request/response-type queries to RAIL and shutdown afterwards, or keep running and keep a connection to the consumer requesting the

query, in the case of a query having requested a change-feed. Their purpose is to first split the posed query into individual queries to the appropriate databases, and aggregating the databases responses in a single message to be delivered as response to the requesting customer.

Load-Balancer, consumer-side.
The load-balancer on the consumer-side does work differently from the load-balancer on the data-provider-side. Where the data is actually passed through the load-balancer on the data-provider side, the load-balancer on the consumer-side serves as a lookup-service, providing the address of the machine currently assigned to instantiate new query-abstraction modules. Similarly to the data-provider-side, the load-balancer shall broadcast its address to allow easy hook-up of consumers.

Management Engine.
The management engine constantly monitors state and availability of all modules, and spawns new modules in case the currently running ones crashed or become unavailable. Further it shall monitor performance metrics of the machines hosting the modules and select appropriate machines to instantiate new data-provider handler and query-abstraction modules on, communicating the new addresses to the load-balancers. The management-engine shall further serve as a central storage for configuration data used by the internal modules.

4   Conclusion

We have drafted the design of a software system capable of representing a dynamic industrial indoor environment, the goal of which is enabling LBS. In addition to the industrial application's requirements we have considered relevant quality measures of software systems such as scalability, robustness and performance and have designed a software architecture that fits these needs. The long term goal with this is to provide an open standard upon which LBS can be developed.

The software system and data modelling have been drafted to meet therequirements outlined in section. The software system addresses important considerations like robustness and scalability. Still, whether the system meets the requirements in production depends on additional factors like the used hardware and infrastructure. In summary, as open and flexible as the system is in its current definition, as much does performance depend on how it will be used. This includes questions like what data will be put into the model and how efficient the concrete algorithms will be. That is why guides for deployment and operations of the system will have to be developed alongside the system itself.

In further work we will proceed to implement the proposed concept to document its viability. This includes developing an accompanying toolset which aids tasks such as e.g. map creation.

120

Acknowledgements

Research was funded under the project IIL by the European Fund for Regional Development (EFRE).

EUROPÄISCHE UNION
Europäischer Fonds für
regionale Entwicklung

References

1. Dhar, S., Varshney, U.: Challenges and business models for mobile location-based services and advertising Commun. ACM **54**(5), 121 (2011). doi: 10.1145/1941487.1941515
2. Xu, Y., Chen, X., Ma, L.: LBS based disaster and emergency management. In: Liu, Y. (ed) 18th International Conference on Geoinformatics, 2010: 18 - 20 June 2010, Beijing, China. 2010 18th International Conference on Geoinformatics, Beijing, China, pp. 1–5. IEEE, Piscataway, NJ (2010). doi: 10.1109/GEOINFORMATICS.2010.5567872
3. Fritsch, L., Scherner, T.: A Multilaterally Secure, Privacy-Friendly Location-Based Service for Disaster Management and Civil Protection. In: Lorenz, P., Dini, P. (eds) Networking - ICN 2005: 4th International Conference on Networking, Reunion Island, France, April 17-21, 2005, Proceedings, Part II. Lecture Notes in Computer Science, vol. 3421, pp. 1130–1137. Springer-Verlag Berlin Heidelberg, Berlin, Heidelberg (2005)
4. Google Public Alerts, https://developers.google.com/public-alerts/, last accessed 2017/11/03.
5. Barnes: Location-Based Services: The State of the Art e-Service Journal **2**(3), 59 (2003). doi: 10.2979/esj.2003.2.3.59
6. Infsoft Whitepaper, https://www.infsoft.de/portals/0/images/solutions/basics/whitepaper/infsoft-whitepaper-de-indoor-positionsbestimmung_download.pdf, last accessed 2017/11/03.
7. IndoorLBS, http://www.indoorlbs.com/, last accessed 2017/11/03.
8. IndoorAtlas, http://www.indooratlas.com/, last accessed 2017/11/03.
9. estimote, https://estimote.com/, last accessed 2017/11/03.
10. Biczok, G., Diez Martinez, S., Jelle, T., Krogstie, J.: Navigating MazeMap: Indoor human mobility, spatio-logical ties and future potential. In: 2014 IEEE International Conference on Pervasive Computing and Communication Workshops (PERCOM WORKSHOPS), pp. 266–271 (2014). doi: 10.1109/PerComW.2014.6815215
11. micello, https://www.micello.com/, last accessed 2017/11/03.
12. Ubisense SmartFactory, https://ubisense.net/de/products/smart-factory, last accessed 2017/11/03.
13. IntraNav, http://intranav.com/, last accessed 2017/11/03.
14. Gunduz, M., Isikdag, U., Basaraner, M.: A REVIEW OF RECENT RESEARCH IN INDOOR MODELLING & MAPPING Int. Arch. Photogramm. Remote Sens. Spatial Inf. Sci. **XLI-B4**, 289–294 (2016). doi: 10.5194/isprsarchives-XLI-B4-289-2016
15. Afyouni, I., Ray, C., Claramunt, C.: Spatial models for context-aware indoor navigation systems: A survey JOSIS(4) (2012). doi: 10.5311/JOSIS.2012.4.73
16. Foote, T.: tf: The transform library. In: 2013 IEEE International Conference on Technologies for Practical Robot Applications (TePRA): 22 - 23 April 2013, Woburn, Massachusetts, USA. 2013 IEEE Conference on Technologies for Practical Robot Applications (TePRA),

Woburn, MA, USA, pp. 1–6. IEEE, Piscataway, NJ (2013). doi: 10.1109/TePRA.2013.6556373

17. Zlatanova, S., Sithole, G., Nakagawa, M., Zhu, Q.: Problems in indoor mapping and modelling Acquisition and Modelling of Indoor and Enclosed Environments 2013, Cape Town, South Africa, 11-13 December 2013, ISPRS Archives Volume XL-4/W4, 2013 (2013)

18. CityGML: Exchange and Storage of Virtual 3D City Models, http://www.citygml.de, last accessed 2017/11/03.

19. GeoJSON, http://geojson.org/, last accessed 2017/11/03.

Comparison of practically applicable mathematical descriptions of orientation and rotation in the three-dimensional Euclidean space

Prof. Dr.-Ing. R. Müller and Dipl.-Wirt.-Ing. (FH) M. Vette M.Eng., A. Kanso M.Sc.

ZeMA - Zentrum für Mechatronik und Automatisierungstechnik gemeinnützige GmbH, Gewerbepark Eschberger Weg 46, Gebäude 9, 66121 Saarbrücken, Germany, a.kanso@zema.de,
WWW home page: http://www.zema.de

Abstract. In the handling technique, the orientation in the three-dimensional Euclidean space is considered or predefined for several assembly tasks. There are different mathematical methods to describe them. Euler presents 24 different conventions that describe any rotation in the space by a rotation about three orthogonal axis. These conventions are singularity-dependent and their deployment is restricted to certain applications and domains. Quaternions overcome the disadvantages of the above mentioned conventions. They are numerically stable and efficient and describe the orientation by a rotation about one axis. This article deals with the comparison of practice-relevant mathematical descriptions.

1 Introduction

If research has widely explored rotation matrice, Euler angles, quaternions and rotation vectors, not one single reference dealing with the study of the practical applicability of mathematical constructs in industrial applications has been founded. Only mathematical principles have been studied and compared in this field.

A freely moving rigid body in the three-dimensional Euclidean space has six degrees of freedom [1], three degrees of freedom for the position in space (x, y, z) and three degrees of freedom for the orientation $(\theta_x, \theta_y, \theta_z)$. In the frame of this article, the orientation discription will be studied. Six degree of freedom robots have the ability to reach any orientation in the three-dimensional Euclidean space. Several industrial robotic applications e.g. welding, insertion and riveting prerequire a specific orientation requirements in order to be implemented.

Robot manufacturers use different conventions to describe the orientation of the robot wrist. The interpolation of the orientation parameter between two given restrictions are greatly dependent on the used convention. This means that if a

© Springer-Verlag GmbH Deutschland, ein Teil von Springer Nature 2018
T. Schüppstuhl et al. (Hrsg.), *Tagungsband des 3. Kongresses Montage*
Handhabung Industrieroboter, https://doi.org/10.1007/978-3-662-56714-2_14

given path has two orientation restrictions on both its ends, a different orientation along it can be achieved when different conventions are used, even though the same path and restriction are processed. A comparison based on simulation results will be implemented in order to discuss this problem. The boundaries of each convention will be examined.

2 Euler angles conventions

The orientation of coordinate system with respect of another frame can be expressed with a 3×3 rotation matrice. A rotation is a displacement in which at least one point of the rigid body remains in its initial position and not all lines in the body remain parallel to their initial orientations [2]. Physically, a rotation can be interpreted as a reorientation of a body without any change in its shape and size [3]. This means that a rotation matrice can rotate a vector $\underline{r} = (x, y, z)^T$ by multiplying the latter with the matrice, while the length of the vector $\underline{r}$ is preserved. Rotation matrices have nine elements while only three parameters are required to define an orientation in space. This implies that six auxiliary relations exist between the elements of the matrice [2]. This redundancy can introduce numerical problems in calculations and often increase the computational cost of an algorithm [4]. Given that rotation matrices are orthogonal, the inverse of the matrice is the transpose of the matrice itself and its determinant is equal to one. Rotation matrice has the following form:

$$D = \begin{pmatrix} b_{11} & b_{12} & b_{13} \\ b_{21} & b_{22} & b_{23} \\ b_{31} & b_{32} & b_{33} \end{pmatrix} \tag{1}$$

Let the matrices D_z, D_y and D_x be the rotation matrices about z, y and z axis respectively, where θ_x, θ_y and θ_z are the angles of the rotation around the above mentioned axis.

$$D_z = \begin{pmatrix} cos\theta_z & -sin\theta_z & 0 \\ sin\theta_z & cos\theta_z & 0 \\ 0 & 0 & 1 \end{pmatrix} D_y = \begin{pmatrix} cos\theta_y & 0 & sin\theta_y \\ 0 & 1 & 0 \\ -sin\theta_y & 0 & cos\theta_y \end{pmatrix} D_x = \begin{pmatrix} 1 & 0 & 0 \\ 0 & cos\theta_x & -sin\theta_x \\ 0 & sin\theta_x & cos\theta_x \end{pmatrix} \tag{2}$$

Euler angles $(\theta_x, \theta_y, \theta_z)$ are most frequently used to represent an orientation of a rigid body in the space. Thus, they are easily applicable and comprehensible [5]. An arbitrary orientation in the space can be described by three rotations in sequence about the coordinate of either a fixed or a moving coordinate frame, such that two consecutive rotations cannot be performed about a parallel axis. If the above mentioned rotations are performed about a fixed coordinate frame, they are therefore designated as extrinsic rotations. Otherwise, if the rotations are performed about axes of local frame, where the latter is modified after each successive rotation, these elemental rotations are designated as intrinsic rotations [6].

The above mentioned sets of Euler angles can be classified in two classes: proper Euler angles and Tait-Bryan angles. The latter can also be found in some literature unter the appellation of cardan angles.

2.1 Proper Euler angles

By proper Euler angles set, the first and the last rotations are about the same axis. That implies twelve sequences of rotations. Six from them are extrinsic (ZX'Z", ZY'Z", YZ'Y", YX'Y", XZ'X" and XY'X"), where superscripts ' and " are used in the case of intrinsic rotations to indicate axes of local frame after the first and second rotation respectively. The other six rotations are intrinsic (ZXZ, ZYZ, YZY, YXY, XZX and XYX). The rotation matrice $D_{ZX'Z''}$ of the rotation sequence ZX'Z" is calculated, where α, β and γ are the rotations angles about Z, X' and Z" respectively.

$$D_{ZX'Z''} = D(Z, \alpha) \cdot D(X', \beta) \cdot D(Z'', \gamma) \tag{3}$$

$$D_{ZX'Z''} = \begin{pmatrix} c_\alpha c_\gamma - s_\alpha c_\beta s_\gamma & -c_\alpha s_\gamma - s_\alpha c_\beta c_\gamma & s_\alpha s_\beta \\ s_\alpha c_\gamma + c_\alpha c_\beta s_\gamma & -s_\alpha s_\gamma + c_\alpha c_\beta c_\gamma & -c_\alpha s_\beta \\ s_\beta s_\gamma & s_\beta c_\gamma & c_\beta \end{pmatrix} \tag{4}$$

Where: $cos(\alpha) = c_\alpha$, $sin(\alpha) = s_\alpha$, $cos(\beta) = c_\beta$, $sin(\beta) = s_\beta$, $cos(\gamma) = c_\gamma$ and $sin(\gamma) = s_\gamma$. If the orientation problem is solved from the opposite direction, this means that $D_{ZX'Z''}$ is the given rotation matrice and the angles α, β and γ are to be calculated. That implies the following solution:

$$\beta = arccos(b_{33}) \quad \alpha = arctan2\left(\frac{b_{13}}{-b_{23}}\right) \quad \gamma = arctan2\left(\frac{b_{31}}{b_{32}}\right) \tag{5}$$

α and γ are not solvable for $\beta = n\pi$ for $n \in \mathbb{Z}$. This implies that proper Euler angles have mathematical singularity in this two positions. This means that this convention can not be used by a robot when the latter has an orientation of $\beta = n\pi$ for $n \in \mathbb{Z}$. Arctan2(y,x) is the inverse of the tangent function that takes into consideration the sign of both y and x to identify the quadrant in which lies the resulting angle.

2.2 Tait-Bryan angles

Tait-Bryan angles, also known as cardan angles, represent a rotation about three different axes. These two different nominations have their roots in the names of mathematicans. Tait-Bryan stands for Peter Guthrie Tait, a 19th-century Scottish mathematical physicist and Cardan comes from the an Italian Renaissance mathematician Gerolamo Cardano. This angles are well known in aerospace engineering and computer graphics [5]. Twelve rotation sequences are classified in this group, six sequences represent extrinsic rotations (ZY'X", ZXY, YZ'X", YX'Z", XZ'Y", XY'Z") and the other six are intrinsic rotations (ZYX, ZXY, YZX, YXZ, XZY, XYZ). The ZY'X" sequence is widely used and is known as Roll, Pitch and Yaw (RPY) convention. The latter is popular in the aeronautics field. Roll, pitch and yaw represent three angles, where roll is the rotation about the Z axis. The direction of the latter is along the positive forward body fixed direction of a vehicle. The Y axis body fixed and directed to the port of the vehicle, the body fixed X axis is directed downwards (see figure: 1). Other literatures consider another arrangement of the X, Y and Z axes. Due to the physical clarity of the RPY convention to extract the angles and axis of rotation, it is commonly used in the robotics community. Most of the robot manufacturers, e.g. KUKA and Universal robots, use this convention to describe the orientation on the robot wrist of the teach pendant. Sometimes though, another convention is used in the controller of the robot in order to calculate the orientation during the path planning. The reason why it is better to use another convention for the controller will be discussed in the upcoming chapters.

The rotation matrice of the RPY convention $D_{RPY'}$ is as follows:

$$D_{RPY} = D(Z, \delta) \cdot D(Y', \eta) \cdot D(X'', \xi) \tag{6}$$

$$D_{RPY} = \begin{pmatrix} c_\delta c_\eta & c_\delta s_\eta s_\xi - s_\delta c_\xi & c_\delta s_\eta c_\xi + s_\delta s_\xi \\ s_\delta c_\eta & s_\delta s_\eta s_\xi + c_\delta c_\xi & s_\delta s_\eta c_\xi - c_\delta s_\xi \\ -s_\eta & c_\eta s_\xi & c_\eta c_\xi \end{pmatrix} \tag{7}$$

Where: $cos(\delta) = c_\delta$, $sin(\delta) = s_\delta$, $cos(\eta) = c_\eta$, $sin(\eta) = s_\eta$, $cos(\xi) = c_\xi$ and $sin(\xi) = s_\xi$. The inverse problem that extracts the rotation angles δ, η and

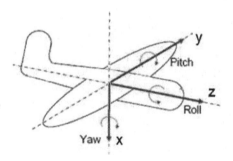

Fig. 1. Roll Pitch Yaw convention.

ξ from the rotation matrice is often of interest. The solution depends on the solving of a set of transcendental equations. This implies:

$$\eta = arcsin(-b_{31}) \quad \delta = arctan2\left(\frac{b_{21}}{b_{11}}\right) \quad \xi = arctan2\left(\frac{b_{32}}{b_{33}}\right) \tag{8}$$

δ and ξ are not solvable for $\eta = \frac{2n+1}{2}\pi$ for $n \in \mathbb{Z}$. This implies that Tait-Bryan angles have a mathematical singularity in this two position. This convention can therefore not be used by a robot when the latter has an orientation of $\eta = \frac{2n+1}{2}\pi$ for $n \in \mathbb{Z}$.

In the previous subsection, 24 Euler angles conventions are listed. Twelve among them represent extrinsic rotations and the other rotations represent intrinsic rotations. It is important to know that due to duality extrinsic rotation sets with the intrinsic rotations, there are effectively only 12 unique parameterizations of a rotation matrice by using successive rotations about principal axes [7]. Analyzing both rotation sequences $D_{ZY'X''}$ and D_{XYZ} implies that they result to the same rotation matrice.

$$D_{ZY'X''} = D(Z, \delta) \cdot D(Y', \eta) \cdot D(X'', \xi) \tag{9}$$

$$D_{XYZ} = D(Z, \delta) \cdot D(Y', \eta) \cdot D(X'', \xi) \tag{10}$$

$$\Rightarrow D_{ZY'X''} = D_{XYZ}$$

It can be derived that mathematical singularity is the main disadvantage of Euler angles convention. In order to physically interpret the singularities, they are said to arise from the gimbal lock situation when two rotational axis align. Physical and practical issues of these convention will be studied in the upcoming chapters.

3 Quaternion

Due to the above mentioned singularities, rotation matrices are numerically unstable [8]. A conventional strategy to overcome these disadvantages to change the representation whenever an object nears to a singularity. These weaknesses in the Euler angles convention led scientists to search for advanced conventions to solve the orientation problem. Unit quaternion provide an intelligent solution to solve this orientation problem. This way, the relevant functions of the quaternion possess no singularity.

The trigonometric form of the complex number of rank 2 $c = a + ib$ can be written as $c = |c|(cos(\theta + isin(\theta))$, wehre θ represents a rotation about an axis that results from cross product of the real and imaginary axes. Hamilton extends complex numbers to a number system of higher (four) spatial dimensions called quaternion, also known as Hamilton quaternion. Quaternion has one real part and three imaginary units i, j and k that make it possible to implement a rotation in the 3D space. William Rowan Hamilton devised the formula of quaternion on the 16 October 1843 with a penknife into the side of the nearby Broom Bridge while he was walking out along the Royal Canal in Dublin with his wife [9]. Let a quaternion be denoted by $\underline{q}$.

$$\mathbf{q} = q_0 + \underline{q} = \left(q_0 \; i \cdot q_1 \; j \cdot q_2 \; k \cdot q_3 \right)^T \tag{11}$$

$$i^2 = j^2 = k^2 = -1 \tag{12}$$

According to the Euler theorem, a rigid body or coordinate frame can be brought from an arbitrary initial orientation to an arbitrary final orientation by a single rotation through the angle of rotation θ about the principal axis $\underline{u}$, also known as Euler vector [10]. Quaternion can be written as:

$$\mathbf{q} = cos\left(\tfrac{\theta}{2}\right) + sin\left(\tfrac{\theta}{2}\right) \underline{u} \tag{13}$$

The norm of quaternion $|\mathbf{q}|$ can be calculated according to the following equation:

$$|\mathbf{q}| = \sqrt{q_0{}^2 + q_1{}^2 + q_2{}^2 + q_3{}^2} \tag{14}$$

Where the norm of unit quaternion is equal to one. Let $\mathbf{p}$ be an arbitrary quaternion, then the product $\mathbf{p} \cdot \mathbf{q}$ can be reduced after applying the formula of the quaternion (see equations above) to:

$$\mathbf{p} \cdot \mathbf{q} = p_0 \cdot q_0 - \underline{p} \cdot \underline{q} + p_0 \cdot \underline{q} + q_0 \cdot \underline{p} + \underline{q} \times \underline{p} \tag{15}$$

For an arbitrary 3D vector $\underline{r} \in \mathbb{R}$ the operator $L_q(\underline{v}) = \mathbf{q}\underline{v}\mathbf{q}^*$ can geometrically rotate $\underline{r}$ through an angle θ about $\underline{q}$. The formulation of $L_q(\underline{v})$ as a matrice rotation results in the following rotation matrice:

$$D_{L_q(\underline{v})} = \begin{pmatrix} 2q_0{}^2 - 1 + 2q_1{}^2 & 2q_1q_2 - 2q_0q_3 & 2q_1q_3 + 2q_0q_2 \\ 2q_1q_2 + 2q_0q_3 & 2q_0{}^2 - 1 + 2q_2{}^2 & 2q_2q_3 - 2q_0q_1 \\ 2q_1q_3 - 2q_0q_2 & 2q_2q_3 + 2q_0q_1 & 2q_0{}^2 - 1 + 2q_3{}^2 \end{pmatrix} \tag{16}$$

Note that $L_q(\underline{v}) = D_{L_q(\underline{v})} \cdot \underline{v}$. Solving the inverse problem to derivate the quaternion parameters from the rotation matrice provides the following equation:

$$\theta = arccos\left(\frac{tr\left(D_{L_q(\underline{v})}\right) - 1}{2}\right) \Rightarrow q_0 = cos\left(\frac{\theta}{2}\right) \tag{17}$$

$$q_1 = \frac{D_{L_q(\underline{v})32} - D_{L_q(\underline{v})23}}{4q_0} \tag{18}$$

$$q_2 = \frac{D_{L_q(\underline{v})13} - D_{L_q(\underline{v})31}}{4q_0} \tag{19}$$

$$q_3 = \frac{D_{L_q(\underline{v})21} - D_{L_q(\underline{v})12}}{4q_0} \tag{20}$$

In the case where $q_0 = 0$ one non-zero quaternion parameter is to be calculated from the diagonal elements of the rotation matrice $D_{L_q(\underline{v})}$) and other parameters can be calculated depending on it. Although quaternion are single free, they are not appreciated from many users in the robotics community. In the case of many users, the four parameters of the quaternion seam not to have a direct physical interpretation, while other conventions such as Euler angles do have one. Developers of handling techniques still prefer them though.

4   Simulation of practical applications

This chapters provides a simulation of the circular path of an industrial robot. The robot must move from the pose A $(x_A \; y_A \; z_A \; \theta_{x_A} \; \theta_{y_A} \; \theta_{z_A})^T$ to the pose B $(x_B \; y_B \; z_B \; \theta_{x_B} \; \theta_{y_B} \; \theta_{z_B})^T$ under the condition that the orientation of its end-effector with respect to the object is conserved during its motion. This can demonstrate a welding application of a circular object. A simulation program is written with the help of Matlab in order to calculate the orientation during

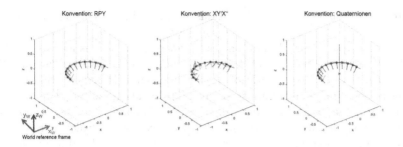

Fig. 2. Simulation representing the orientation based on RPY, XY'X" Proper Euler angles convention and Quaternion.

the desired motion of the robot. The circular object lies on the XY-plane. The orientation of A and B is defined according to the RPY convention as ($R = 0°, P = -45°, R = -85°$) and ($R = -180°, P = -45°, R = -85°$) respectively, where A and B are located opposite to each other. Let N be the number of increments that must be simulated along the path. The orientation parameters at each increment are linearly interpolated. For example θ_{x_n} is calculated as following:

$$\theta_{x_n} = \theta_{x_A} + \frac{n}{N}(\theta_{x_B} - \theta_{x_A}) \tag{21}$$

Where $n \in \mathbb{N}, n <= N$ θ_{y_n} and θ_{z_n} are calculated analogous to θ_{x_n}. The path is simulated according to RPY, XY'X" proper Euler angles convention and quaternion (see figure: 2). This shows that when RPY and quaternion are used the orientation along the path is conserved. Numerical calculation of the orientation at each increment assures this result. On the other hand, the desired orientation is not achieved when the XY'X" Proper Euler angles convention is used. The comparison between the orientation at A and B shows that, since the object lies in the XY-plane, the orientation on the pose B can be achieved by a rotation about the Z axis starting from A. XY'X" convention expresses this rotation by a sequence of rotations about the X, Y', and X" axes. Thus the desired orientation at the increment is not achieved using this convention. The same object is now

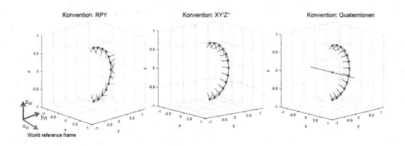

Fig. 3. Simulation representing the orientation based on RPY, XY'X" Proper Euler angles convention and quaternion.

located at YZ-plane and the orientations are predefined according the to XY'Z" as $(\theta_{x_A} = 0°, \theta_{y_B} = 45°, \theta_{z_B} = 85°)$ and $(\theta_{x_B} = 180°, \theta_{y_B} = 45°, \theta_{z_B} = 85°)$ respectively (see figure: 3). This shows that the orientation is lost when RPY is used since it represents here a rotation about the X axis. Quaternion and XY'Z" convention conserve the orientation at the increments. This shows that the Euler angles provide different orientations when interpolating between two predefined boundaries. The orientation towards an object is conserved when the rotation along the path occurs about the first rotation axis of the Euler angles set convention. Quaternion represents a general solution to solve the above mentionned problem, since they express any rotation in space with a rotation about one axis.

5 Conclusion

This article shows that the quaternion gives a practical and numerical stable solution to solve the orientation problem of non-linear path in space. Euler angles solutions are numerically unstable and are not accurate when integrating the incremental changes of orientation over time. The simulation above shows that Euler angles can provide a solution to conserve the orientation along a circular path if and only if the object lies on a plane spanned from the X, Y or Z axes. Beside the fact that quaternion are singularity free, this article shows that they are also practically more efficient than other conventions.

References

1. VDI Rechtlinie 2861 Blatt 1 : Montage- und Handhabungstechnik; Kenngrößen für Industrieroboter; Achsbezeichnungen.Düsseldorf, (1988).
2. Siciliano, B., Khatib, O. (Hrsg.): Handbook of Robotics. Berlin, Heidelberg: Springer (2008).
3. Brannon, R.M.: A review of useful theorems involving proper orthogonal matrices referenced to three dimensional physical space. http://www.mech.utah.edu/brannon/public/rotation.pdf(2002)
4. Funda, J., Russell H. T., Richard P.P.: On homogeneous transforms, quaternions, and computational efficiency. IEEE Transactions on Robotics and Automation, 6(3), pp. 382388 (1990).
5. Diebel, J.: Representing Attitude: Euler Angles, Unit Quaternions, and Rotation Vectors. Stanford University. Stanford, California 94301-9010 (2006).
6. Vidakovic, J. Z., Lazarevic, M. P., Kvrgic, V. M., Dancuo, Z. Z., Ferenc, G. Z.: Advanced Quaternion Forward Kinematics Algorithm Including Overview of Different Methods for Robot Kinematics. University of Belgrade. 198 VOL. 42, No 3 (2014).
7. Craig, J. J.: Introduction to Robotics Mechanics and Control 3rd edition. Pearson pp. 3947 (2005).
8. Corke, P.: Robotics, Vision and Control: Fundamental Algorithms in MATLAB. Springer Tracts in Advanced Robotics, Vol. 73, Springer Verlag Berlin Heidelberg (2011).
9. Kuipers, J. B.: Quaternions and Rotation Sequence. New Jersey (1999).
10. Tomasi, C.: Vector Representation of Rotations. https://www.cs.duke.edu/courses/fall13/compsci527/notes/rodrigues.pdf(2013)

"Human-In-The-Loop"- Virtual Commissioning of Human-Robot Collaboration Systems

Maximilian Metzner[1,a], Jochen Bönig[2], Andreas Blank[1], Eike Schäffer[1]
and Jörg Franke[1]

[1] Institute for Factory Automation and Production Systems (FAPS)
University Erlangen-Nuremberg (FAU), Erlangen, Germany
[2] SIEMENS AG Digital Factory Division GWE, Erlangen, Germany
[a]maximilian.metzner@faps.fau.de

Abstract. Human-robot collaboration (HRC) has the potential to increase the degree of automation, and thus productivity, throughout many industries. However, complex safety considerations and a lack of appropriate planning tools still prohibit a more widespread application. In this paper, the use of virtual commissioning (VC), an established tool for the validation of common automated systems, for the validation of HRC-systems is proposed. For this, the structure of a traditional VC environment is combined with a digital human model (DHM). To ensure adequate behavioral fidelity, a real-time overlay of human action and the virtual automated system, using motion capture technology, is imperative. This also requires the live visualization of the simulation environment via virtual reality (VR). The structure of such a simulation system is presented and evaluated. Furthermore, the cost and benefit for this new method is contrasted.

Keywords: Human-Robot Collaboration, Virtual Commissioning, Digital Human Modeling, Motion Capturing.

1 Introduction

Assembly is a major cost driver in industrial production [1]. Enhanced customer focus and shortened technology lifecycles have led to an increasing number of product variants and shorter product lifecycles [2]. Common fully automated assembly solutions, focused on high production quantities and short cycle times, cannot meet consequential requirements regarding flexibility and mutability [1]. Manual processes, especially in handling and assembly tasks impair profitability and productivity, especially in high-wage countries [3]. Hybrid assembly systems, often enabled through HRC, offer a compromise between the two paradigms [4]. To fully realize potential benefits of HRC assembly systems, comprehensive planning methods and tools need to be established [5]. This paper aims to provide production planners with a tool for the virtual planning and validation of HRC production systems, increasing worker acceptance and planning quality. For this, a short overview of the current state of the art, and its deficiencies, is given. Afterwards, the structure of the proposed system is presented and its features are discussed. The paper closes with an outlook on further research and implications.

© Springer-Verlag GmbH Deutschland, ein Teil von Springer Nature 2018
T. Schüppstuhl et al. (Hrsg.), *Tagungsband des 3. Kongresses Montage*
Handhabung Industrieroboter, https://doi.org/10.1007/978-3-662-56714-2_15

2 State of the Art

In the following chapter, a short overview of the state of the art in HRC, VC and DHM is given. Concluding the chapter, deficiencies in the state of the art are deduced.

2.1 Human-Robot Collaboration

First, a short introduction to HRC and associated research is given. HRC aims to combine the specific characteristics of manual and automated processes by abolishing the strict separation of workspaces for human workers and robots. It also enables hybrid production systems, where specific tasks can be automated, e.g. for ergonomic reasons, with little to no additional space requirements. Four forms of HRC can be identified: coexistence, sequential cooperation, parallel cooperation and collaboration (see Fig. 1).

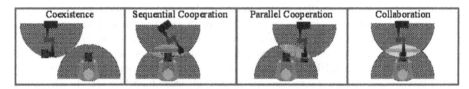

Fig. 1. Characteristic forms of HRC (see also [5])

HRC applications need to fulfill the same normative requirements as common robot applications, most notably the ISO 10218 norm [6; 7]. However, due to potential overlaps of workspaces, the absence of physical barriers and even permissible contact scenarios between worker and robot, safety considerations are generally more complex for HRC applications. Specific requirements for HRC systems are detailed in the ISO/TS 15066 specification, e.g. acceptable force and pressure limits for human-robot-collisions considering standard operation and reasonably foreseeable misuse [8]. Still, concerns about safety and liability are cited as major barriers for the widespread application of HRC. [5; 9] Addressing them is one objective of the presented approach.

To increase planning quality and realize efficiency potentials, virtual planning of HRC systems is in scientific focus. One key factor for the virtual validation of any HRC system is the virtual representation of the interacting human(s) [10; 11]. Recent studies have focused on integrating ergonomic simulations, with an offline human model, into the planning tools [12]. In some cases, MC technologies have been used to animate the human model to later use in the simulation [13]. Further research is being conducted on determining physical conditions, such as impact forces, into virtual HRC simulation to allow for more in-depth safety considerations [14].

2.2 Virtual Commissioning

Since virtual validation for HRC is not yet sophisticated, a look at fully automated systems, were virtual validation is common, is worthwhile. VC is a method to validate the real-to-life behavior of an automated system, especially focusing on the interactions with the programmable logic controller (PLC), before physical commissioning. The actual PLC code is used to operate a simulation via its outputs that, in turn, generates the input signals for the PLC. [15] Two types of VC can be distinguished: hardware-in-the-loop (HiL) and software-in-the-loop (SiL). While HiL uses an actual PLC to evaluate inputs generated by the simulation and generate outputs operating the simulation, SiL uses a simulated PLC. While HiL thus requires the simulation to be real-time, it can offer further insight into controller behavior rather than just system behavior. [16]

The main benefits of VC are an increase in planning quality, simulation of dangerous or damaging scenarios, reduction of software design flaws as well as a shorter commissioning phase due to parallelization of VC with the product development process. However, as VC requires a holistic virtual model of the entire relevant system, additional efforts for virtual modeling are generated. [15; 17]

2.3 Digital Human Modeling and Human-in-the-Loop Simulations

As humans are an integral part of every HRC system, the digital modelling of humans is of great relevance to virtual HRC validation. DHM originated in North America in the context of air and space research. First applications were representations of humans as objects, e.g. in crash simulations [18]. DHM is currently used, among other applications, for ergonomic analysis of workplaces. These simulations can either be run with offline human models, performing predetermined motions, or online, using motion-capture technology. Certain simulations require the human to be 'in the loop', reacting to simulated inputs. These "human-in-the-loop" (HUIL) simulations are predominately used for applications like driving or flight simulators. [19; 20]

A key aspect to these simulations is the concept of behavioral fidelity, ensuring that the behavior during the simulation reflects real-life behavior. This is, for instance, supported by deeper immersion, e.g. in the sensory domain. [19; 21]

2.4 Deficiencies

While there is a variety of established methods and tools for the design and evaluation of both manual as well as automated workplaces, there is not yet a comprehensive toolkit for HRC systems.

Recent scientific research in the field of HRC, focusing on the representation of humans in virtual planning environments, has primarily relied on offline human simulation, leaving human behavior predefined. This constraint, however, is critical for the evaluation of HRC systems, in which the human is an active participant in the automated process [19]. Additionally, predefined behavior blocks hardly meet behavioral fidelity requirements necessary to accurately simulate a HRC system, e.g. when considering misuse, which is in scope of safety considerations described above [8].

In contrast to common automated systems, classic virtual commissioning, linking a PLC with a purely virtual simulation model, cannot sufficiently represent the behavior of a HRC system. VC, however, could potentially help to validate safety features in perilous situations and decrease safety validation and commissioning duration.

The suggested approach expands mentioned HRC validation capabilities beyond the state of the art by enabling realistic user interaction, increasing behavior fidelity and thus ultimately increasing planning quality while shortening physical commissioning and safety validation time.

3   Human-in-the-Loop Virtual Commissioning System

To further increase planning efficiency and help identifying crucial safety issues early on, a comprehensive virtual planning method for HRC systems is developed. In the following chapter the general requirements, approach and the components of the system are displayed along with the structure of the suggested system (see Fig. 2).

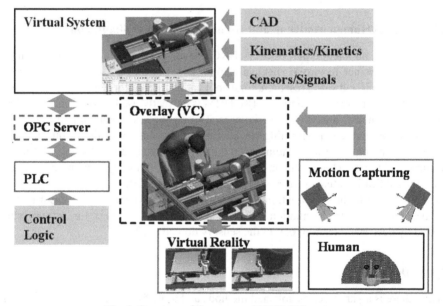

Fig. 2. Structure of the proposed HUIL VC system

As stated above, the key factor for enhancing the virtual validation of HRC systems is to increase its applicability for safety considerations and control logic testing. To realize this, an increase regarding the behavioral fidelity of the human involved in the simulation is necessary. This is achieved by an enhanced immersion into the simulation, creating more realistic human feedback, and an online mapping of this feedback back into the virtual system. Safety aspects to be investigated include collision zones, -velocity and -geometries, squeezing hazards, sensor placement and overall safety functionality. Non-safety aspects focus on workflow training, ergonomics evaluation and cycle time optimization.

For the virtual system, all relevant components have to be assembled in Computer Aided Design (CAD). Their kinematic as well as logical behavior needs to be defined.

In a last step, the control logic, including all signals from the virtual overlay, has to be programmed on the PLC. The PLC can either be simulated (SiL) or physical (HiL).

Immersion is increased by the use of VR, stereoscopically visualizing the virtual system for the human operator. A MC system tracks the operator's movement and maps them into the virtual overlay, in turn influencing the simulation. For the system to properly interact with the human operator and PLC, the simulation has to run in real-time, with visualization along with human capturing and mapping being executed at minimal latency.

To ensure industrial applicability, the implementation is planned using components compliant with industrial standards. The virtual system is designed with Siemens NX 11® CAD software, making use of its built-in library functions to reduce modeling efforts for recurring applications. The simulation environment runs on Siemens Process Simulate 13.1.2®. CAD data exchange uses the JT format, while kinematic/kinetic information is stored in XML. HiL VC runs on a Siemens S7-1500® series PLC coupled with a Siemens Simulation Unit® as interface. Stereoscopic VR is generated in the Process Simulate® environment and displayed using a nVisor ST50® head-mounted display. In a first series, motion capturing will be realized with a Microsoft Kinect V1® sensor. Depending on the achieved results, more elaborated tracking equipment, for larger detection zones, finger tracking and improved accuracy, will be evaluated.

4   Discussion

The proposed method is focused on improving the virtual planning quality of HRC systems, emphasizing safety considerations. The key concept is an online overlay of virtual system and real human behavior. Requirements include a complete virtual model of the system, a VR and MC system, as well as hard- and software for common VC. The benefits of this approach include a reduction of physical commissioning time, considering both PLC validation and HRC safety testing (see Fig. 3).

Setting up a complete virtual model of the system requires significant effort, and is also often a barrier for the implementation of VC in common automated systems [15]. As this method is however designed to further enhance virtual planning of HRC systems towards VC, it is assumed that a virtual model, representing major parts of the system, e.g. for process/kinematics validation or offline programming, is already available. Generating PLC code for VC is no additional effort in itself, since only the due date for its validation is changed. In this case, the additional effort for the proposed approach is thus limited and overcompensated by a reduction in physical commissioning time and cost, e.g. for later system design changes. This is especially beneficial for the introduction of new variants into a running system, a trend fueled by challenges presented in the introduction, or for implementing process optimizations.

The hard- and software for VC, VR and MC presents a sizeable monetary investment, especially considering that the simulation has to be run in real-time. However, since the equipment can be reused for future applications, lowering the cost per application, this investment becomes profitable for system integrators or larger companies, wishing to implement multiple HRC applications.

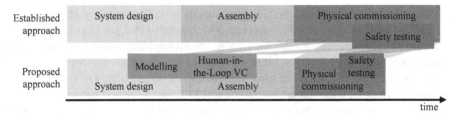

Fig. 3. Efforts and benefits of HUIL VC (see also [22])

Additional to safety considerations, the proposed HUIL approach can be used for training and workshop purposes. With the virtual overlay, workers can safely train their working tasks with a virtual robot model behaving realistically, even before the physical assembly of their work station. The reaction of the robot to every conceivable scenario can be experienced by the worker first-hand. This benefits employee satisfaction and increases acceptance of the HRC system, which is crucial for the successful implementation of HRC [5]. On a side note, the repeated simulation of the working cycle can also be used to verify planned cycle times of the system in scope. Accurate cycle times are useful for value stream design and system performance optimization [23].

5   Summary and Outlook

In this paper, the need for enhanced virtual planning of HRC applications is defined. The state of the art is described in the fields of VC, HRC and DHM. Benefits of VC, as well as definitions and restrictions for HRC are detailed. Also, experiences from the field of DHM considering behavioral fidelity and HUIL simulations are portrayed. VC for HRC systems, as a combination of aforementioned methods, to ensure adequate behavioral fidelity, is identified as a potential field of action for improving achievable planning quality through virtual tools. Requirements, components and the structure of a proposed approach are detailed. The paper is then concluded by discussion on the economic feasibility for different-sized companies and the benefits expected to be achieved. It is inferred that the method is best suited for companies specialized in the development of HRC systems or that plan on implementing multiple applications with available know-how in common VC and robot simulation.

In a next step, the proposed approach will be evaluated using a real-life use-case considering a HRC system in power electronics production. This will offer further insight into the practical feasibility and achievable benefits in the virtual planning and validation of HRC systems as well as the match of the simulation model with reality.

In the future, due to rapid technological developments in the consumer segment, especially regarding VR/MC-systems, and related cost degression, this validation method will become more suitable for a broader segment of companies, facilitating the planning and validation of HRC and ultimately helping to realize its potential.

References

1. LOTTER, B.; WIENDAHL, H.-P.: *Montage in der industriellen Produktion: Ein Handbuch für die Praxis*. 2. Aufl. 2013. Berlin, Heidelberg: Springer Berlin Heidelberg; Imprint; Springer, 2012 (VDI-Buch)

2. WESTKÄMPER, E.: *Modulare Produkte - modulare Montage*. In: *wt Werkstattstechnik online* (2001), Nr. 91, S. 479–482. http://werkstattstechnik.de/wt/article.php?data%5Barticle_id%5D=304

3. FELDMANN, K.; SCHÖPPNER, V.; SPUR, G.: *Handbuch Fügen, Handhaben und Montieren*. München: Carl Hanser Verlag GmbH & Co. KG, 2013

4. BEUMELBURG, K.: *Fähigkeitsorientierte Montageablaufplanung in der direkten Mensch-Roboter-Kooperation*. Heimsheim: Jost-Jetter, 2005 (IPA-IAO-Forschung und -Praxis 413)

5. BENDER, M.; BRAUN, M.; RALLY, P.; SCHOLTZ, O.: *Leichtbauroboter in der manuellen Montage - einfach einfach anfangen: Erste Erfahrungen von Anwenderunternehmen*. Stuttgart, 2016

6. DIN EN ISO 10218 - 1:01.2012, *Industrieroboter – Sicherheitsanforderungen – Teil 1: Roboter*

7. DIN EN ISO 10218 - 2:06.2012, *Industrieroboter – Sicherheitsanforderungen – Teil 2: Robotersysteme und Integration*

8. DIN ISO/TS 15066:04.2017, *Roboter und Robotikgeräte - Kollaborierende Roboter (ISO/TS 15066:2016)*

9. Institut für angewandte Arbeitswissenschaft e.V. (ifaa): *Mensch-Roboter-Kollaboration - Zahlen | Daten | Fakten*. https://www.arbeitswissenschaft.net/fileadmin/user_upload/Downloads/Factsheet_MRK_6.pdf – Überprüfungsdatum 2017-10-06

10. DEUSE, J.; BUSCH, F.; Weisner, Kirsten, Steffen, Marlies: Gestaltung soziotechnischer Arbeitssysteme für Industrie 4.0. In: HIRSCH-KREINSEN, H. (Hrsg.): *Digitalisierung industrieller Arbeit: Die Vision Industrie 4.0 und ihre sozialen Herausforderungen*. Baden-Baden: Nomos, 2015 (edition sigma).

11. BÖNIG, J.; FISCHER, C.; BROSSOG, M.; BITTNER, M., et al.: Virtual Validation of the Manual Assembly of a Power Electronic Unit via Motion Capturing Connected with a Simulation Tool Using a Human Model. In: ABRAMOVICI, M.; STARK, R. (Hrsg.): *Smart Product Engineering: Proceedings of the 23rd CIRP Design Conference, Bochum, Germany, March 11th - 13th, 2013*. Berlin, Heidelberg: Springer Berlin Heidelberg, 2013, S. 463–472

12. BUSCH, F.: *Ein Konzept zur Abbildung des Menschen in der Offline-Programmierung und Simulation von Mensch-Roboter-Kollaborationen*. 1.

Auflage. Herzogenrath: Shaker, 2016 (Schriftenreihe Industrial Engineering 17)

13. BUSCH, F.; WISCHNIEWSKI, S.; DEUSE, J.: Application of a character animation SDK to design ergonomic human-robot-collaboration. In: *Proceedings of the 2nd International Symposium on Digital Human Modeling (DHM)*, 2013, S. 1–7

14. HYPKI, A.: *Kollaborative Montagesysteme: Verrichtungsbasierte, digitale Planung und Integration in variable Produktionsszenarien (3.* Workshop "Mensch-Roboter-Zusammenarbeit"). Dortmund, 29.03.2017

15. WÜNSCH, G.: *Methoden für die virtuelle Inbetriebnahme automatisierter Produktionssysteme.* München: Utz, 2008 (IWB-Forschungsberichte 215)

16. VERMAAK, H.; NIEMANN, J.: Virtual commissioning: A tool to ensure effective system integration. In: *Proceedings of the 2017 IEEE International Workshop of Electronics, Control, Measurement, Signals and their Application to Mechatronics (ECMSM): May 24-26, 2017, Donostia-San Sebastian, Spain.* [Piscataway, NJ], [Piscataway, NJ]: IEEE, 2017, S. 1–6

17. LIU, Z.; SUCHOLD, N.; DIEDRICH, C.: Virtual Commissioning of Automated Systems. In: ADENIJI, D. O. (Hrsg.): *Optimization of IPV6 over 802.16e Wi-MAX Network Using Policy Based Routing Protocol:* INTECH Open Access Publisher, 2012

18. DUFFY, V. G.: *Handbook of digital human modeling: Research for applied ergonomics and human factors engineering.* Boca Raton: CRC Press, op. 2009

19. LOPER, M. L.: *Modeling and Simulation in the Systems Engineering Life Cycle: Core Concepts and Accompanying Lectures.* London: Springer London, 2015 (Simulation Foundations, Methods and Applications)

20. BÖNIG, J.; PERRET, J.; FISCHER, C.; WECKEND, H., et al.: Creating realistic human model motion by hybrid motion capturing interfaced with the digital environment. In: *International FAIM Conference,* 2014, S. 317–324

21. BÖNIG, J.; FISCHER, C.; WECKEND, H.; DÖBEREINER, F.; FRANKE, J.: *Accuracy and Immersion Improvement of Hybrid Motion Capture based Real Time Virtual Validation.* In: *Procedia CIRP* 21 (2014), S. 294–299

22. ZÄH, M. F.; WÜNSCH, G.: *Schnelle Inbetriebnahme von Produktionssystemen.* In: *Werkstattstechnik* 95 (2005), Nr. 9, S. 699–704

23. ABELE, E.; WOLFF, M.; MANZ, A.: *Optimierung von Wertströmen.* In: *ZWF Zeitschrift für wirtschaftlichen Fabrikbetrieb* 107 (2012), Nr. 4, S. 212–216

Simulation-based Verification with Experimentable Digital Twins in Virtual Testbeds

Ulrich Dahmen and Jürgen Roßmann

Institute for Man-Machine-Interaction, RWTH Aachen University,
Ahornstraße 55, 52074 Aachen, Germany
dahmen@mmi.rwth-aachen.de, rossmann@mmi.rwth-aachen.de
WWW home page: http://www.mmi.rwth-aachen.de/

Abstract. As modern systems become more and more complex, their design and realization, as well as the effective management of engineering projects evolve to increasingly challenging tasks. This explains the continuous need for appropriate cross-domain methodologies in order to handle complexity and thereby create reliable systems. One important aspect is the substantiation that a specific system design is suitable for its intended use, which is usually achieved by testing. Unfortunately, those tests are carried out after the system has been produced so that the elimination of possible errors and defects causes high efforts and costs. This paper introduces a systematic approach for a simulation-based verification and validation support by using experimentable digital twins during the entire product life cycle. It allows to test the system under development in various virtual scenarios before it is implemented and tested in reality and thus reduces the risk of lately detected system design errors which increases the reliability of the development process.

Keywords: Systems Engineering, Modeling and Simulation, Virtual Testbed, V-Model, Verification and Validation, Digital Twin

1 Introduction

Technical systems usually are very complex due to the large number of different subsystems with diverse relationships and dependencies, the increasing number of internal states and the high degree of dynamic behavior [1]. It is characteristic for those systems that small errors can cause major consequences including the failure of the entire system. A quite popular example is the loss of the Mars Climate Orbiter (MCO) in 1999 in cause of an overlooked type mismatch of metric and imperial units inside the navigation component causing a wrong course correction. Therefore, the orbiter mistakenly approached Mars up to 57 km instead of 150 km and was destroyed by the significant higher atmospheric pressure at this height [2].

© Springer-Verlag GmbH Deutschland, ein Teil von Springer Nature 2018
T. Schüppstuhl et al. (Hrsg.), *Tagungsband des 3. Kongresses Montage*
Handhabung Industrieroboter, https://doi.org/10.1007/978-3-662-56714-2_16

Following the current trends of digitalization and interconnection as discussed in the context of "Industry 4.0" and "Internet of Things" combined with the continuing miniaturization and integration of different technologies into so-called "Smart Systems", the inherent complexity of modern systems increases considerably. Consequently, the system's development process as well as the qualification process become more and more complex, too. System failures like the aforementioned are usually very dangerous and cause high costs. For the design, production, and timely marketing, it is therefore necessary to understand all individual components, subsystems, and the overall system at all stages of the development process in theory and in practice. To manage this major challenge, appropriate engineering processes have been established. Combining those with appropriate simulation techniques allows to compare and evaluate different system designs and to reduce the number of physical prototypes.

This paper introduces a systematic approach for the realization of "Experimentable Digital Twins" (EDT) and their integration into systems engineering processes to support both the design and verification of the system by using state of the art simulation techniques at every stage of the development process. Chapter 2 outlines the principles of Digital Twins and Systems Engineering as a basis for the subsequent work. Chapter 3 focuses on the concept of Experimentable Digital Twins, followed by Chapter 4 that shows a systematic integration into the engineering process leading to the concept of Simulation-based Verification and Validation. Finally, Chapter 5 gives an application example from a space project followed by a conclusion and outline for future work in Chapter 6.

2 State of the Art

Most modern Systems are designed modular in both software and hardware. That means that the functionality and correct behavior of the system depend on the correct and purposeful interaction of modules with a local intelligence. Consequently the analysis of the overall system in its operational environment becomes more and more important. Methods from the traditional Systems Engineering face this complexity problem by appropriate process models that include detailed specifications of all components as well as the overall system with all relevant interfaces and relationships. A very popular and approved engineering procedure model is the traditional **V-Model** [3], that starts with a hierarchical top-down analysis and design phase, followed by the implementation and reverse bottom-up integration and test phase. It is a chronological (timely structured) and stage-oriented (logical structured) model that directly defines appropriate test cases for every design step and thereby intrinsically integrates quality management right from the beginning. However, the V-Model has a crucial problem: Possible errors and drawbacks in the system design usually are not found before the corresponding validation activities *at the end* of the system development, causing high efforts for changes, adjustments, and optimization.

Another challenge results from the fact that the development of modern systems requires the integration of a multitude of discipline domains causing the need to create interfaces and coordinate between standard methods of the involved disciplines [4]. To manage this challenge the methodology of **Model-based Systems Engineering** (MBSE) starts playing an increasing role in Systems Engineering. The main aspect is the "formalized application of modeling to support system requirements, design, analysis, verification and validation activities beginning in the conceptual design phase and continuing throughout development and later life cycle phases" [5]. As a model-centric approach this methodology focuses on creating digital information repositories that carry and interconnect all relevant information of the system and thereby provide a common information base.

This leads to the current approach to mirror all relevant aspects of a real-life system (also called technical asset[1]) into a virtual substitute summarized by the terms "Digital Shadow" (original "Digitaler Schatten") and "Digital Twin". The **Digital Shadow** is one of the central cornerstones of Industry 4.0 and describes a sufficient accurate image of the processes in production, development or other nearby areas with the aim to create a complete data base with all relevant data [7]. Similarly, the **Digital Twin** describes a virtual image of a real subject (human, application) or object (machine, environment) that reflects all relevant static and dynamic properties (based on [8]).

3 Experimentable Digital Twins for Simulation-based Verification

One focus of this paper is the introduction of an implementation approach that transforms the abstract concept of a Digital Twin to a concrete implementation semantics based on modeling and simulation techniques, leading to the term "Experimentable Digital Twin". But before discussing this it is necessary to look at the term "Model" in detail to prevent linguistic confusion. Models in general are simplified representations of a real or hypothetical scenario that only include those aspects that are relevant to the issue under consideration with a given tolerance. [9], [10]. This paper distinguishes between two main classes of models, on the one hand the *Information Model*, and on the other hand the *Simulation Model*. The latter comprises the executable code to run a specific simulation including the corresponding conceptual, mathematical, and computerized model and data. The Information Model instead describes the information repository as created by the MBSE-methodology including cross-connected system models for requirements, behavior, structure, etc. Figure 1 illustrates the different types of model and their primal interaction.

---

[1] A technical asset or technical item is an artifact produced especially to fulfill a role within a system [6].

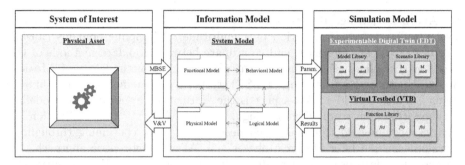

Fig. 1: Classification and basic structure of Experimentable Digital Twins

The **Experimentable Digital Twin** (EDT) of a specific System of Interest (SoI) as defined here is a collection of model elements (submodels of the Simulation Model) forming the EDT's *Model Library*, combined with a collection of simulation scenarios collected in the ETD's *Scenario Library*. The Model Library has a hierarchical structure that mirrors the system's physical structure (comparable to a mechanical parts list) and contains models for the overall system, assemblies, components, etc. In contrast the Scenario Library contains specific simulation experiments, which are comprehensive simulation models build up from the model elements of the EDT's Model Library and perhaps other external (global) model libraries (e.g. for modeling the operational environment or the interaction with other systems). These simulation scenarios can be executed ("running the simulation") in a **Virtual Testbed** (VTB)[2] that provides the runtime environment for the EDT. The VTB implements the *Function Library* which contains all relevant simulation functions (algorithms) that can be applied to the EDT (data) for gathering simulation results. None of the three libraries are static. In fact they evolve during the project representing the current progress. Usually the Model Library and the Function Library contain several implementations for the same element to represent different levels of detail. In the same way different simulation scenarios of the Scenario Library may represent different levels of integration.

The presented definition of an EDT enables to "execute" the System Model meaning that the system specification becomes "experimentable". Thus, it is possible to test and demonstrate the performance of newly developed systems under different boundary conditions as well as the interactions between components and subsystems in various operational scenarios.

---

[2] Virtual Testbeds (VTB) provide a methodology for a comprehensive and cross-application simulation environment that allows the analysis of an overall system in its envisaged operational environment, covering all relevant aspects in the system's life cycle. Therefore, a Virtual Testbed is equipped with efficient simulation algorithms ranging from rigid body dynamics and soil mechanics up to the simulation of various actuators and sensors, combined with up to date Virtual Reality and 3D simulation techniques [11], [12].

4 Integration into Systems Engineering

As already mentioned the problem with the traditional V-Model for systems engineering is the risk of late detected design flaws. The integration of EDTs into this process model leads to a systematic use of simulation techniques enabling a simulation-based verification and validation of the system right from the beginning of the development. As shown in Figure 2, the area between the descending and ascending branch is filled up with a novel virtual branch which transfers all stages of the ascending branch to a virtual test level that can be applied as early as possible. This way the virtual branch closes the timely gap between system design and system test. One central point of this methodology is the concept of

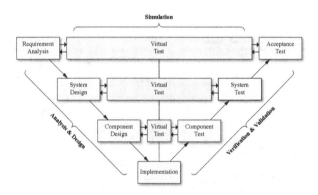

Fig. 2: V-Model extended by a virtual branch

simulation-based verification and validation. Therefore it is important to have a common understanding of "verification" and "validation" (V&V), because there are many slightly different understandings of these terms resulting from the wide range of fields of application (refer to [13], [14], [15], and [16]) and the common language meaning that is unfortunately imprecise. It is important to realize that there are different views on which the terms "verification" and "validation" can be applied to. The first one is the traditional modeling and simulation view, where V&V should prove the correctness of gathered simulation results. On the other hand is the systems engineering view with the aim to prove the correctness of a developed system. For the concept of simulation-based verification and validation it is necessary to distinguish between those views. Therefore, this paper defines and uses the following terms:

- *System*-Verification (SVer): Examination, whether a *system* has been implemented correctly. In practice: Does the implemented system comply with its specification?
- *Model*-Verification (MVer): Examination, whether a (simulation-) *model* has been implemented correctly. In practice: Does the executable model represent the conceptual model correctly?

- *System*-Validation (SVal): Examination, whether the right *system* has been developed. In practice: Does the designed system fulfill all user requirements?
- *Model*-Validation (MVal): Examination, whether the right *model* has been simulated. In practice: Does the executable model represent the real system well enough for the intended use?

The central point of this definition is the distinction between *System*-V&V and *Model*-V&V that is necessary because simulations with adequate models (Digital Twins) should be used for the verification and validation of the system. Figure 3 illustrates this in detail. The design phases provide **Simulation-based System-Verification and -Validation** activities by performing *virtual* tests with the EDT, based on the test specification of the previous phase.

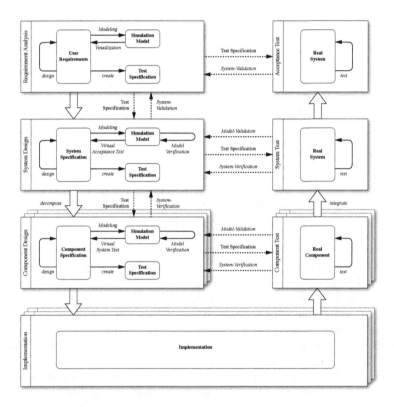

Fig. 3: Use of simulation models as virtual prototypes for Simulation-based Verification and Validation during the development process

This way the use of EDTs extends the traditional horizontal V&V by a vertical V&V. Once the entire system has been designed, specified and virtually verified at all stages, the system can be implemented, integrated and tested in reality. The test results finally can be used for a Model-Validation of the EDT.

5   Application Example

A first application example is the upcoming iSAT-1 space mission that combines a tracking payload with an iBOSS-based[3] satellite platform to form a space segment, which is intended to extend the operational small object tracking service from the international cooperative for animal research using space (ICARUS). Since modular satellite systems are very complex due to the number of partly autonomous subsystems an EDT of the iSAT-1 was implemented during the initial ECSS Phase 0 study[4] to simulate and investigate the early system designs in different operational environments. Figure 4 exemplary shows an excerpt of the satellite's structural decomposition as part of the system model, as well as specific simulation models from the EDT's Model Library.

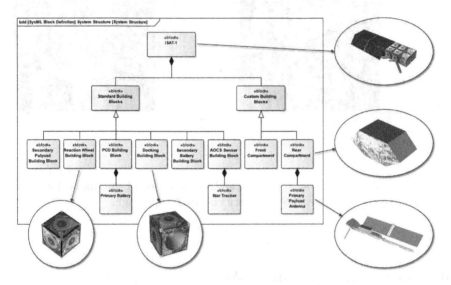

Fig. 4: Structural decomposition of the modular satellite (excerpt) and associated elements in the EDT's Model Library

One focus of the mission analysis was the determination of the optimal mission orbit since the selection of a specific orbit directly influences the satellite's thermal budget and earth surface coverage time, a key property for the tracking service that should be provided by the iSAT-1. Thus, a scenario was defined us-

---

[3] The iBOSS approach is a DLR-funded initiative to develop and bring into operational use a novel modular spacecraft concept enabling On-Orbit Servicing (OOS) and On-Orbit Assembly (OOA), respectively reconfiguration and expansion [17].

[4] The ECSS (European Cooperation for Space Standardization) is an initiative established to develop a coherent set of standards for use in all European space activities. Those standards divide the life cycle of space projects into 7 phases, beginning with a so-called "Phase 0", which aims to define and analyze the overall mission [18].

ing the advanced rendering technique and the rigid body dynamics engine from the Virtual Testbed as simulation functions. Both are applied with a fast forward scheduling to simulate several hours of the satellite's orbital movement in minutes. Figure 5 illustrates on the left side the fully assembled virtual prototype of the system under development, and on the right side results of the scenario simulation. The darker areas at the earth's surface mean that the satellite does not have covered these areas so far.

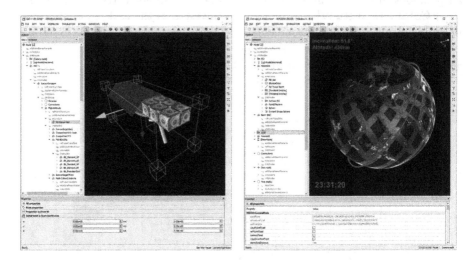

Fig. 5: Assembled overall system (left) and visualization of the simulation results for the analysis of the satellite's surface coverage time (right)

6 Conclusion and Future Work

This paper introduces an engineering procedure model based on the V-Model and extended by a virtual prototype approach to verify and validate a system under development with the help of Experimentable Digital Twins. This enables vertical Simulation-based Verification and Validation activities before the system is implemented and tested. This helps to reduce the risk of lately detected and thus very expensive system design errors. One of the next steps is the development of formal methods that support the applicability of the EDT including the generation of simulation models, the Model-V&V as well as the exchange with tools and workflows of the MBSE methodology.

Acknowledgments

This work is part of the project "iBOSS-3", supported by the German Aerospace Center (DLR) with funds of the German Federal Ministry of Economics and Technology (BMWi), support code 50 RA 1504.

References

1. H. Ulrich and G. J. B. Probst, *Anleitung zum ganzheitlichen Denken und Handeln.* Bern: P. Haupt, 1991.
2. A. G. Sephenson, "Mars climate orbiter: Mishap investigation board phase i report," tech. rep., Mishap Investigation Board, Nov. 1999.
3. K. Forsberg and H. Mooz, "The relationship of system engineering to the project cycle," in *Proceedings of the 12th Internet World Congress on Project Management,* 1994.
4. T. Weilkiens, *Systems Engineering with SysML/UML.* Elsevier Science & Technology, 2008.
5. International Council on Systems Engineering, "Systems engineering vision 2020," Tech. Rep. INCOSE-TP-2004-004-02, INCOSE, 2007.
6. VDI/VDE-Gesellschaft Mess- und Automatisierungstechnik: Fachausschuss Industrie 4.0, "Industrie 4.0 - technical assets," tech. rep., Verein Deutscher Ingenieure e.V., Düsseldorf, 2015.
7. Wissenschaftliche Gesellschaft für Produktionstechnik WGP e.V., "WGP-Standpunkt Industrie 4.0," tech. rep., WGP e.V., 2015.
8. M. Grieves, "Digital twin: Manufacturing excellence through virtual factory replication: A white paper," 2014.
9. National Academy of Science and Engineering, and Communication Promoters Group of the Industry-Science Research Alliance, "Recommendations for implementing the strategic initiative industire 4.0," tech. rep., acatech and CPG, Frankfurt/Main, 2013.
10. VDI 3633 Blatt 1, "Simulation of systems in materials handling, logistics and production," 2014.
11. J. Rossmann and M. Schluse, "Virtual robotic testbeds: A foundation for e-robotics in space, in industry—and in the woods," in *Proceedings of the 4th International Conference on DeSE,* (Dubai), pp. 496–501, 6th - 8th December 2011.
12. J. Rossmann, M. Schluse, C. Schlette, and R. Waspe, "Control by 3d simulation a new erobotics approach to control design in automation," in *Proceedings of the 5th ICIRA,* vol. Part II of *LNAI 7507,* pp. 186–197, Montreal, Quebec, Canada: Springer, October 3-5, 2012 2012.
13. SCS Technical Committee on Model Credibility, "Terminology for model credibility," *SIMULATION,* vol. 32, no. 3, pp. 103–104, 1979.
14. W. L. Oberkampf and C. J. Roy, *Verification and validation in scientific computing.* Cambridge: Cambridge University Press, 2010.
15. Department of Defense (DoD), *DoD Directive No. 5000.59: Modeling and Simulation (M&S) Management,* 1994.
16. American Society of Mechanical Engineers, *Guide for verification and validation in computational solid mechanics : ASME V&V 10-2006.* New York: ASME, 2006.
17. J. Weise, K. Brieß, A. Adomeit, H.-G. Reimerdes, M. Göller, and R. Dillmann, "An intelligent building blocks concept for on-orbit-satellite servcing," in *Proceedings of the International Symposium on Artificial Intelligence, Robotics and Automation in Space (iSAIRAS),* 2012.
18. ECSS-M-ST-10C Rev. 1, "Space project management - project planning and implementation," 2009.

A Flexible Framework to Model and Evaluate Factory Control Systems in Virtual Testbeds

Tim Delbrügger[1] and Jürgen Roßmann[2]

[1] Department of Robot Technology, RIF e.V.
tim.delbruegger@rt.rif-ev.de
[2] Institute for Man-Machine-Interaction, RWTH Aachen
rossmann@mmi.rwth-aachen.de

Abstract. Modern decentralized factory control systems promise to beat traditional hierarchical systems with regard to flexibility and fault tolerance. Despite the benefits, many companies hesitate to make the switch because they have little experience with decentralized systems and cannot fathom the specific implications for their production scenario. To support companies in evaluating specific solutions, we recommend to use Virtual Testbeds, which provide close-to-reality multi-domain simulations and interactive 3D visualizations of production systems. In order to represent different control strategies in Virtual Testbeds, we propose a new framework in which control strategies consist of modular control units. By mixing these control units in different ways, centralized, decentralized and hybrid approaches can easily be implemented. We evaluate the framework by building and controlling a digital twin of a real production system. Together with the unique features of Virtual Testbeds, the proposed framework is an essential step towards a case-by-case decision support for factory control decentralization projects.

Keywords: Virtual Testbeds, factory automation, decentralized factory control, control modelling

1 Introduction

Consider a factory in the automotive sector in which the owning company wants to evaluate the changes that a modern, decentralized or hybrid factory control system would impose on the shop floor. Since such a big transformation poses a considerable risk, the company needs a simulation with detailed evaluation and interaction possibilities in order to build trust into some control system candidate.

As frequently updating multiple models for different simulation domains is very expensive, a single simulation model should be usable for prototyping the new control system as well as for approximations of costs, virtual commissioning and virtual reality presentations of the factory.

Most of these wishes are common across today's factories. Especially finding the optimal degree of control decentralization for a specific production scenario

© Springer-Verlag GmbH Deutschland, ein Teil von Springer Nature 2018
T. Schüppstuhl et al. (Hrsg.), *Tagungsband des 3. Kongresses Montage*
Handhabung Industrieroboter, https://doi.org/10.1007/978-3-662-56714-2_17

is far from trivial. In order to evaluate decentralized or hybrid production systems, many studies like [1] employ discrete-event simulations. However, these simulations operate on a high level of abstraction, where many possible problems in the physical world are not considered at all. Virtual Testbeds follow the opposite approach and include detailed digital twins of the real system, its (simulated) behavior and the relevant environment. Based on this, they can additionally contain discrete-event simulations and agent-based simulations inside the same model. All in all, Virtual Testbeds seem to be a perfect match for the requirements of the scenarios mentioned above.

Our framework for modeling factory control systems is especially inspired by the "Strategy" pattern from object-oriented software, as presented in [2]. This pattern enables us to build a factory model that is largely independent of the used control system, and thus lets users of the system switch easily between different control strategies.

Throughout the paper, we will refer to a part of an airbag factory with the needs mentioned above. The production processes involve a warehouse, eleven injection molding, two painting and three assembly lines. The plastic airbag covers are injection molded and then painted, before they are combined with other, preassembled parts in the assembly lines.

2 State of the Art

This section gives a short overview of previous contributions to Virtual Testbeds and the relation between control theory and optimization.

2.1 Virtual Testbeds

Virtual Testbeds are good candidates for factory automation because they are of proven value for many technologies that form the foundation of the Industry 4.0 initiative. Some examples are the use for mobile robotics [3] and the development of sensor-enabled applications [4]. Specific industrial applications include the evaluation of autonomous transport vehicles [5], the development of novel user interfaces for reconfigurable industrial assembly [6], navigation based on Building Information Models [7] and the evaluation of industrial robots with regard to energy consumption[8].

2.2 Decision supportControl and Optimization

Tasks in the real world can be interpreted as optimization problems because the circumstances typically define the conditions and objectives of the problem. Consider an industrial transport going to some destination: most probably, the costs (including fuel, personnel, taxes, fees, etc.) should be minimized while reaching the destination in time. By explicitly formulating the task as an optimization problem, we enable algorithms to understand them. The formulation as a problem also yields a separation of the control task and the strategy to solve it.

The idea to model control tasks as optimization problems is very common in the literature. The area of optimal control derives control strategies that are optimal with regard to the optimization problem [9]. And if no optimal control strategy is available, metaheuristics can achieve good solutions [10].

3 Modeling Factory Control Systems

Any factory control system takes high-level production orders and translates these into specific actions of all kinds of actors in the factory. To be able to create production plans, information about the capabilities of the production systems must be present in the proposed framework. Because of this, each station has a set of production mappings it can perform. Each production mapping contains a set of material input and material output.

3.1 Entities and Tasks

The Virtual Testbed contains two main types of objects: those that are physical entities in the real factory and those that are not physical (like control logic). An overview of some important entities is given in fig. 1. The control logic for the entities is implemented via solvers. These solvers can be seen as a replacement for the traditional controllers in factory automation, as they solve control problems of known entities.

In our framework, each control task is stored at the corresponding entity. For example, a transport unit appends path finding queries to its set of current problems, instead of just solving it directly.

Decoupling tasks from their solvers is important because it allows to switch more easily between centralized and decentralized control structures, as the tasks are the same in both cases, but the location and linkage of solvers is different. A completely hierarchical approach to transport vehicle control can introduce a global path planner, which plans collision-free paths for all vehicles in the factory. The same path planning problems can be solved in an autonomous way by letting the vehicles plan their own paths and introducing more sophisticated avoidance behaviors to eliminate collisions.

Self evidently, the general idea of decoupling tasks and solvers is not new, as it is a central part of mathematical optimization and the "Strategy" design pattern in object-oriented software. However, this paper advances this general idea by letting a solver dynamically group problems and solve them together.

Fig. 2 shows another simple example, where the factory gets the task to produce three cars. The entities of these three products are created with the need for a production plan. Following the trend of mass customization, we always create production plans for individual products, because this allows full flexibility in the scheduling, while mass production of the same products is still possible.

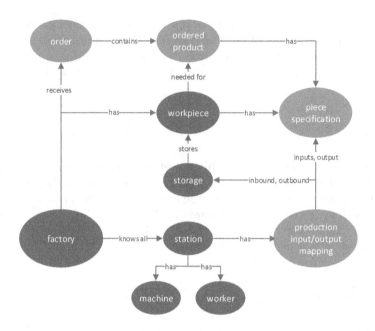

Fig. 1. Overview of some important entities (ellipses) in the presented framework and their relations (green arrows). Entities that have a body and are visible in the Virtual Testbed are painted blue, while purely logical entities are grey. Solvers can relate to all entities and are not shown.

3.2 Solvers

Solvers are software components that compute solutions to the control tasks. They can use any means, like analytical optimization, specialized algorithms, heuristics or genetic algorithms. In order to apply the internal methods, solvers include a mapping between the factory's semantic model and the internal problem model. Linked to each solver is a list of entities whose problems it solves.

Some control tasks in the factory might be solved as optimization problems. Deviating from the common mathematical description of optimization problems, we model the control task's optimization objective as a property of the solver. This makes sense because many solvers are not able to optimize different objectives. Especially heuristics are often directly tuned for a special purpose. On the other hand, generic multi-objective genetic algorithms try to find pareto-optimal solutions [11]. By defining separate goals for each solver, we are able to express this variability in the model. We also gain the ability to combine a complex and sometimes slow multi-objective algorithm with a simpler, faster, single-objective algorithm as a fallback.

Until now, the control tasks are just defined. But how are they solved and where is the solver located? This refers to the important research field of decentralization in factory control, which is why our framework leaves an experimenter in the Virtual Testbed total freedom to choose between a completely decentralized

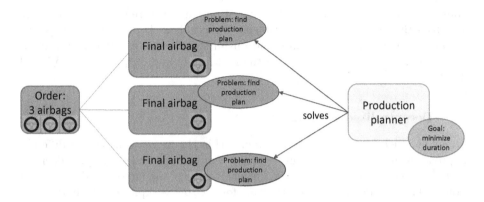

Fig. 2. Entities (blue) can have control problems attached (red). Solvers like a production planner can find solutions to these with regard to some goals (green). The production planner in this example finds all the production plans by minimizing the total production duration. In a more decentralized version, each product might have its own solver attached.

approach with local ad-hoc decisions and a hierarchical top-down planning. As we see in the next sections, both are possible and can be mixed.

3.3 Big problems vs small problems

Some factory control problems are complex in the sense that they contain lots of smaller problems. For example, finding a scheduling for all stations contains the single station scheduling problems. Combining all the problems into one problem has the benefit that a global optimum for the whole factory can be obtained, instead of just optimal solutions for individual machines. However, in our framework we have decoupled the definition of the problem and the solving, which means that even if all the single problems are defined separately, a solver can still decide to combine them.

All in all, we gain flexibility without reducing solution quality by defining the individual, small problems instead of the complex ones.

3.4 Hierarchic control approach

In the traditional factory control pyramid, human experts or specialized software systems plan production in an MES system, which then distributes tasks to Supervisory Control and Data Acquisition (SCADA) and Programmable Logic Controller (PLC). Automation in this pyramid is done in hand-crafted program code, which has very limited flexibility.

The proposed framework can easily represent such a hierarchic approach by introducing just one solver for all similar problems inside one context. For example, all stations of a factory can have the same scheduler associated. Robots, conveyors and other actors inside the station can get a reference to the same solver for their motion planning problems, mimicking a classic PLC. As a specific solver can have arbitrary inputs and outputs, it is also possible to represent a traditional PLC component with the real inputs and outputs in the Virtual Testbed.

3.5 Decentralized control

To define a decentralized control system within the framework, several different approaches are possible and can be mixed. For example, there can be "decision nodes" in the factory with sensors, knowledge about the factory and processing power. These nodes can solve problems for nearby workpieces, like finding appropriate stations for the next production steps or finding a nearby transport vehicle.

Another idea is to virtually attach solvers to the same factory entities as the problems. For example, a station can solve its own scheduling problems. The ordered product can route its work-in-progress pieces through the factory and allocate time slots on workstations. Obviously, this virtual attachment is no problem in the Virtual Testbed.

Regardless of whether the control system is hierarchic, decentralized or a hybrid, we want to provide a thorough evaluation to be able to compare the performances of different control systems, which leads to the next section.

4 Implementation and Demonstration

As a starting point to evaluate a specific factory configuration, we chose to measure and plot throughput time, stock size and machine utilization. Logistic experts can use these to follow "Logistic Operating Curves Theory" of Nyhuis and Wiendahl [12]. There are lots of other key performance indicators that can be computed during a factory's operation and different factory operators often have divergent ideas about their importance. Therefore, a user of the Virtual Testbed can combine a diverse set of performance measurements as needed for the specific case via a visual programming framework [13].

Besides automatically gathered measurements, a real-time 3D visualization of the simulations allows engineers to gain insights into many aspects and implications that are not measured. It dramatically eases the interdisciplinary communication, for example in the case that the building needs to be changed as part of an adaption scenario.

In order to evaluate the presented framework, we built a Virtual Testbed that represents the factory part mentioned previously. Fig. 3 shows a visualization of the painting stations with the injection molding machines in the background. A small fleet of autonomous vehicles transports material between the stations, while some operation statistics for the painting stations are shown.

To demonstrate the modular concept of the framework, we employed several simple heuristics, which can be replaced in the future. For example, our scheduling problem solver finds shortest paths through the production network with respect to available time slots across known stations with a modified A* algorithm. Due to its general implementation, it can be used as a central planner for all orders and stations as well as a decentralized planner for single orders and a subset of stations. The association of transport orders and vehicles is solved in a decentralized way by letting the transporters pick the nearest task whenever they are available.

For the evaluation of factory control logic, a realistic set of orders is important. Because of this, we based our simulation on an anonymized, real dataset of one exemplary shift from the factory. As the choice of any single time period might introduce a bias, we recommend to analyze several different production scenarios before making a final decision. In order to enable the user of the Virtual Testbed – usually a production or logistics engineer – to quickly switch between different control concepts, we added an input to all solvers which can enable or disable them. Then the solvers that should cooperate in one control strategy can be connected to the same value or logic so that they are active at the same time.

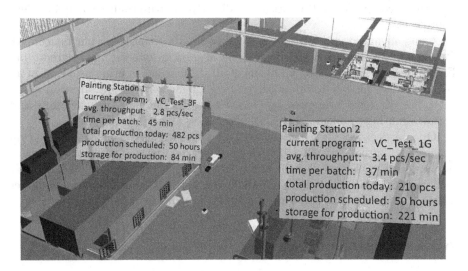

Fig. 3. Visualization in the Virtual Testbed with live performance data

Conveyors and transport vehicles transport material between the stations. Vehicles use the path planning and following framework presented in [7], which incorporates the building information from a Building Information Model and the dynamic interior of the factory.

While the simulation is running, the system shows live statistics for each station, transport unit and the whole factory. Using the visual programming framework, all sensor data and statistics can also be exported to CSV files for external analysis.

5 Conclusion

We presented a new framework to model and evaluate factory control systems in Virtual Testbeds. Because of its modularity, engineers can mix hierarchic and decentralized control elements to achieve an optimal control strategy for their specific production system. For the mentioned part of the exemplary factory used through the paper, the proof of concept implementation of the framework is finished, so that the detailed evaluation of various control systems can start. The system is ready to model and evaluate various centralized, decentralized or hybrid factory control systems in a Virtual Testbed.

6 Outlook

In the future the proposed framework will be used to develop new solvers and their combinations in order to find the best control strategies for specific production scenarios. Utilizing the framework's ability to also model hierarchical control structures, companies and scientists can directly compare traditional and modern approaches.

Using advanced optimization-based solvers as a scenario-specific benchmark for modern decentralized approaches will also be an interesting research topic. For example, the scheduling problems of all stations and routing problems of all vehicles in the factory could be solved by a single complex optimization-based solver like the ones presented by Gamst for the combined job scheduling and network routing problem [14].

Once a decision for a factory control approach has been made based on experiments in the Virtual Testbed, there are several ways to transfer the approach into production. As one idea to omit a costly re-implementation, Roßmann et al. demonstrated how real sensor data and a connection to real actors can empower Virtual Testbed software to control complex systems in the real world [15].

Acknowledgements

This work is supported by the German Research Foundation (DFG) as Tim Delbrügger is a member of the Research Training Group 2193.

References

1. Scholz-Reiter, B., Freitag, M., Beer, C.D., Jagalski, T.: Modelling and analysis of autonomous shop floor control. In: 38th CIRP Int. Sem. on Manufacturing Systems – CIRP ISMS '05, Florianopolis. pp. 16–18 (2005)
2. Gamma, E., Helm, R., Johnson, R., Vlissides, J.: Design Patterns: Elements of Reusable Object-oriented Software. Addison-Wesley Longman Publishing Co., Inc., Boston, MA, USA (1995)

3. Rossmann, J., Jung, T.J., Rast, M.: Developing virtual testbeds for mobile robotic applications in the woods and on the moon. In: 2010 IEEE/RSJ International Conference on Intelligent Robots and Systems. pp. 4952–4957. IEEE (2010)
4. Thieling, J., Roßmann, J.: Virtual testbeds for the development of sensor-enabled applications. In: Schüppstuhl, T., Franke, J., Tracht, K. (eds.) Tagungsband des 2. Kongresses Montage Handhabung Industrieroboter, pp. 23–32. Springer Berlin Heidelberg, Berlin, Heidelberg (2017)
5. Rossmann, J., ten Hompel, M., Eilers, K.: Ermittlung der Leistungsverfügbarkeit zellularer Intralogistiksysteme mit Hilfe von Simulations- und VR-Techniken, https://www.logistics-journal.de/proceedings/2014/4059
6. Schlette, C., Kaigom, E.G., Losch, D., Grinshpun, G., Emde, M., Waspe, R., Wantia, N., Roßmann, J.: 3d simulation-based user interfaces for a highly-reconfigurable industrial assembly cell. In: 2016 IEEE 21st International Conference on Emerging Technologies and Factory Automation (ETFA). pp. 1–6. IEEE (2016)
7. Delbrügger, T., Lenz, L.T., Losch, D., Rossmann, J.: A navigation framework for digital twins of factories based on building information modeling. In: 2017 22nd IEEE International Conference on Emerging Technologies and Factory Automation (ETFA). pp. 1–4. IEEE (2017)
8. E. Kaigom, M. Priggemeyer, J. Rossmann: 3d advanced simulation approach to address energy consumption issues of robot manipulators – an eRobotics approach. In: Proceedings for the joint conference of ISR 2014, 45th International Symposium on Robotics, Robotik 2014, 8th German Conference on Robotics. VDE-Verl., Berlin and Offenbach (2014), http://ieeexplore.ieee.org/document/6840208/
9. Craven, B.D.: Control and Optimization. Chapman & Hall Mathematics Series, Springer US, Boston, MA and s.l. (1995), http://dx.doi.org/10.1007/978-1-4899-7226-2
10. Siarry, P. (ed.): Metaheuristics. Springer, Cham (2016)
11. von Lücken, C., Barán, B., Brizuela, C.: A survey on multi-objective evolutionary algorithms for many-objective problems. Computational Optimization and Applications 25(4), 536 (2014)
12. Nyhuis, P., Wiendahl, H.P.: Fundamentals of Production Logistics: Theory, Tools and Applications. Springer-Verlag, Berlin, Heidelberg (2009), http://dx.doi.org/10.1007/978-3-540-34211-3
13. Schlette, C., Losch, D., Rossmann, J.: A visual programming framework for complex robotic systems in micro-optical assembly. In: Proceedings for the joint International Conference of ISR/ROBOTIK 2014 - 45th International Symposium on Robotics (ISR 2014) and the 8th German Conference on Robotics (ROBOTIK 2014). pp. 750–755. VDE Verlag Berlin, Munich (2014)
14. Gamst, M.: Exact and heuristic solution approaches for the integrated job scheduling and constrained network routing problem. Combinatorial Optimization 164, Part 1, 121–137 (2014)
15. Rossmann, J., Schluse, M., Schlette, C., Waspe, R.: Control by 3d simulation – a new eRobotics approach to control design in automation. In: Hutchison, D., Kanade, T., Kittler, J., Kleinberg, J.M., Mattern, F., Mitchell, J.C., Naor, M., Nierstrasz, O., Pandu Rangan, C., Steffen, B., Sudan, M., Terzopoulos, D., Tygar, D., Vardi, M.Y., Weikum, G., Su, C.Y., Rakheja, S., Liu, H. (eds.) Intelligent Robotics and Applications, Lecture Notes in Computer Science, vol. 7507, pp. 186–197. Springer Berlin Heidelberg, Berlin, Heidelberg (2012)

System architecture and conception of a standardized robot configurator based on microservices

Eike Schäffer[1,a], Tobias Pownuk[1], Joonas Walberer[1], Andreas Fischer[2], Jan-Peter Schulz[3], Marco Kleinschnitz[4], Matthias Bartelt[5], Bernd Kuhlenkötter[5], Jörg Franke[1]

[1] Institute for Factory Automation and Production Systems (FAPS),
University Erlangen-Nuremberg, http://faps.fau.de
[a]Eike.Schaeffer@faps.fau.de
[2] Robert Bosch GmbH, AndreasFabian.Fischer@de.bosch.com
[3] ICARUS Consulting GmbH, jan-peter.schulz@icarus-consult.de
[4] Infosim GmbH & Co. KG, kleinschnitz@infosim.net
[5] Institute of Production Systems (LPS), Ruhr-University of Bochum (RUB)

Abstract. The design and integration of robotic-based automation solutions is a common problem for robotic component providers and especially for their consumers. In this work, a standardized robot configurator is introduced, based on a modular system architecture and best-practice solutions. Starting with a minimum viable prototype providing an intuitive web-based configurator, customized robot applications can easily be planned, visualized, simulated and finally realized. The presented robotic configurator is based on microservice architecture, which is a modern, scalable and complexity-reducing solution for the overall system. This paper demonstrates how an exemplary configuration process could be handled to get an impression about the prospective use of pre-configured robotic solutions.

Keywords: Configurators, robot configuration, microservice architecture, minimum viable prototype

1 Introduction

The increasing possibilities and functionalities of robot systems and the continuously growing technical requirements in industrial applications (e.g. high precision, reproducibility and safety) lead to high levels of complexity for the successful implementation of robotic automation solutions. Robotics engineers need a deep understanding of robot kinematics, programming and a comprehensive experience to meet the specific consumers' requirements. Since necessary engineering tools are not available or are isolated applications, the functional and secure integration of individual peripheral components is often realized in expensive trial-and-error procedures. The development processes for robot solutions are therefore characterized by high engineering costs for robotic system engineers and integrators. For this reason, individual, special-purpose

© Springer-Verlag GmbH Deutschland, ein Teil von Springer Nature 2018
T. Schüppstuhl et al. (Hrsg.), *Tagungsband des 3. Kongresses Montage Handhabung Industrieroboter*, https://doi.org/10.1007/978-3-662-56714-2_18

and therefore complex and unique solutions are created. However, their engineering expenses do not achieve an economical cost-benefit ratio, especially for small and medium-sized enterprises.

To overcome this issue, we propose a configurator for robot-based production systems. An appropriate concept is presented within this paper. By utilizing standardization and reuse of software, hardware, and peripheral components a significant reduction in quotation, complexity and engineering expenses for robotic applications will be possible. The aimed solution will be a modular platform for planning and simulation of robot-based systems. The underlying microservice architecture allows a systematic development, application, and marketing in the areas of industry, logistics, and services.

A major benefit of the proposed solution is the needlessness of expert knowledge while planning a robot-based production system. In order to achieve this goal, the platform includes a robotic configurator that suggests appropriate best-practice robot component combinations, depending on process and interface characteristics. By using a standardized engineering tool with unified interfaces for kinematics, effectors, sensors, peripherals, and controls as well as publicly available software libraries and knowledge-bases, a rapid implementation of robotic automation concepts will be achieved. The complexity of the planning process is reduced by decreasing the number of options for the corresponding automation task.

This article is structured as follows. Section 2 provides a comprehensive insight into the state of the art and the challenges of the implementation of robotics solutions. The system architecture for a standardized configuration process is presented in Section 3. Section 4 describes an exemplary configuration procedure of a robot system based on best-practice solutions. A summary and an outlook about further research work is given in Section 5.

2 State of the art / related work

2.1 Planning and simulation of robot-based production systems

Looking at the software tools available today in the field of industrial robotics, they are increasingly focusing on the simulation of automation concepts as a system integration aid. Known simulation and planning software, such as Microsoft Robotics Studio [1], Siemens Process Simulate [2], RoboLogix [3], or Webots [4], enable the design and verification of robot applications in dynamic 3D environments and thus provide an early detection of process-related issues. Also, approaches to work simultaneously with these tools exist, e.g. the solution developed within the conexing research project [5]. Those tools possess a high complexity due to the large number of provided options and they can hardly be used by unexperienced persons. Instead, that complexity should be reduced while still receiving a good solution.. The demand for design methodologies and concepts to provide the best-practices of reconfigurable robotic systems is becoming increasingly important in the industrial automation field [6]. The design process is often a poorly defined, complex and iterative procedure, where the needs and specifications of the required artifact are not refined until the design is almost completed [7].

Especially in this early product development step, it is crucial, that designers are provided with an interactive and vision-oriented tool, which supports in the decision-making process to build up reconfigurable robotic automation solutions. Currently available tools (e.g. robolink, [8]), however, only focus on a certain task, while platforms that cover the entire product development process from design phase to virtual commissioning are required.

Furthermore, a transfer of a once developed concept for an automation task to other applications is limited. As a result, the elaboration of a robotic production system is often carried out by intensive manual exchanges of information between providers and integrators. The use of web-based technologies represents one of the most powerful tools for sharing, pooling and distributing information between project's team members [9]. By using web-based platforms, it is possible to carry out a simultaneous and collaborative design processes with an efficient transfer of knowledge between the project partners [7]. In [10], the optimization of plant engineering and the efficient operation of complex production facilities based on community systems, using knowledge management techniques, were investigated. Other approaches regard a web-based worker information system, where data streams from different controller databases are coordinated and ergonomically presented to the operator [11, 12]. These projects have shown that a web-based cockpit, which dynamically generates an overview of upcoming tasks, has a positive effect on the interaction between automation systems and humans [13].

2.2 Microservice architecture

During a lecture in 2006, the CTO of Amazon, Werner Vogels, spoke about small teams that develop and operate services with their own databases. This was the first time that microservices were mentioned [14]. Even though there is no official definition, a microservice architecture is a service-oriented architecture in which software is made up of small independent services. These services are very loosely coupled, small and focused on a single feature of the software [15]. There are four principles for the development of microservices: a service has only one task, can be programmed within two weeks, services work together, and only universal interfaces may exist between them [14]. However, the design of the overall architecture may take considerably longer.

Microservices usually implement the concepts of distributed systems and service-oriented architectures more consistently than any other system, so that the potential of modular services is better used [16]. Also, developers are forced to modularize the system, because each services is independent and only connected to other services via universal interfaces. Due to the low dependency between the microservices, each one can be replaced with little effort. The result is a system that can be updated with minimal resources and is not permanently bound to individual programming languages or systems such as a specific databases. Straightforward updating is a competitive advantage, because it ensures that the platform is always up to date and enables the constant use of modern software libraries and frameworks based on the latest standards [16]. Because of the modularization, each service can be independently scaled. By using a microservice oriented architecture, the effort to develop and keeping a system up to date is reduced, which is of course a competitive advantage [14].

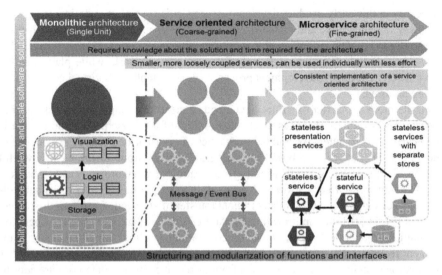

Fig. 1. A microservice approach represents a modern and scalable solution, with a simultaneous complexity reduction of the overall system (after [17]).

Figure 1 sketches the differences between the architecture approaches. In contrast to monoliths, there is a technological freedom for each individual service. This offers two major advantages. New technologies can be tested on a microservice without affecting other services. In use, the potential of the technology can be better determined. The risk of utilizing a new technology is low because it can be exchanged at any time. The second major advantage of the technological freedom is the ability to use the most suitable tool for each task. Databases are the best example. For monoliths, it is necessary to use the same database with the same properties for each task. However, it might be best to use different database types depending on the task. For example, a data store for documents is the good fitting database type to save the master data of a platform. However, this is not suitable for the relationship between robot components where a graphical database should be used. Both database types are therefore required for the creation of a robot configurator. [16]

3 Standardized robot configurator

In order to enable non-experts the planning and simulation of robot-based production systems, we are creating a web-platform providing a standardized robot configurator. The platform will guide the user from its automation task to a running production system. Hence, it will comprise tools from different domains, e.g. layout creation, virtual commissioning, or order placement. The various sub-tasks are realized as microservices, as sketched in Figure 2. To reduce the time to market, we define a minimum viable prototype that already provide a solution to a specified task. However, the individualization of the solution is limited. Following services are available in the minimum viable prototype:

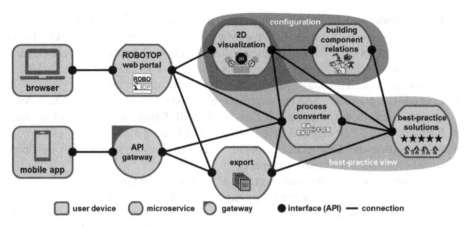

Fig. 2. Microservice architecture of the minimum viable prototype.

- **Process converter:** Convert process input into robotics configuration. Based on simple user questions, a start scenario is selected for the configuration. This step reduces complexity because the user only has to adapt single components instead of building an entire system. For example, a user would like to select the scenario "pick and place" with a range of one meter and a specific gripping component.
- **Best-practice solutions:** The best solution for a scenario or process with fixed variables is shown to the user. With the aid of the service "best-practice solutions", the most suitable components will be selected based on the input of the process converter.
- **2D visualization:** Icons and images are used to display the composition of the components to the user. Also, this service enables the user to change the configuration.
- **Building component relations:** This service is responsible for determining the possible composition of the individual components. For this, e.g. a graph-based database is useful, in which all components with possible relationships and restraints are listed.
- **Export:** Export service is responsible for exporting data. The data includes bill of materials (BOM) lists. This is the last service, which is one of the minimum requirements.

In order to provide a broad spectrum of functionalities, further services will be developed. E.g. these services offer a sophisticated visualization and modification of the current state of the corresponding planned production system. Depending of the knowledge of the user, they allow to influence the outcome of the platform, i.e. the design of the production system suitable for the task specified by the user. Following services are considered for extending the platform:

- **3D visualization:** This service will complement the tasks of 2D visualization. It visualizes the scenario in 3D using for example WebGL. Because the development of a 3D environment is more tedious, it is faster to start development with 2D visualization.

- **Robot simulation:** With robot simulation, a path planning for the robots can be created. At this time only robots with an existing controller library are supported. However, this covers most commonly used industrial robots like articulated robots, palletizers, or SCARA. This task also consists of further services like the manipulation of 3D objects, the creation of path locations and other applications.
- **Component provider:** Service to provide a component's information. The information is described by using e.g. the AutomationML format or another roof format. It may be used for storage and exchange of robot planning data.
- **Sub-components:** This service is mainly a database, which shows other possible components to the selected components.
- **One-Stop-Shop:** The one-stop-shop lists all the data about the components.
- **User-data:** A dedicated service for managing the data of the users.
- **Rights management:** Every user has specific rights, which control what content can be seen, read or altered by him. These rights are managed by this service.
- **Payment system:** High security is required for the payment. Therefore, the payment system is usually outsourced to specialized companies. To ensure that the external workers cannot change the source code, a service is also suitable here.

4   Robot configuration process

The following section describes a hypothetical configuration process. This example is not meant to give a statement about the usability, the subjective perceived effort, or intuitiveness, but to give in impression about the workflow. The exemplary configuration process is shown in Figure 3 and starts with the scenario query for a required robot solution out of a general production process input. This is one way to reduce the solution space and the complexity. The various robot solutions with the greatest potential are presented in primary categories, which are subdivided into further subcategories. By now, the user can select between the primary categories "pick and place", "packaging" and "palletizing" as well as "quality control". The best-practice default scenario based on a default bill of material with a spatial arrangement is presented to the user and can be reconfigured in detail by means of partial configurations. The minimum viable configuration process is based on a layout and component specification as shown in Figure 3. As a part of the layout specification, the user can redefine sub-configurations to determine the actual positions. In addition, the basic requirements for the component (weight, size, shape and surface) can be adapted by the user. Based on these preconditions, it is possible to select a gripper that meets the component's frame conditions and then a robot that meets the payload and reach requirements. The configuration solution is stored in a standard text readable exchange format (e.g. XML or JSON) and is ready for export. The basic configurator offers a simple and fast overview of the most suitable automation solutions for the specific applications to the user, based on fact that only one click is required to display a first scenario. The start scenario is then successively adapted to the real scenario. The preselection of a potential solution generated in the configuration process can be shared with colleagues and system integrators simply via a web link, which promotes an internal and external company collaboration.

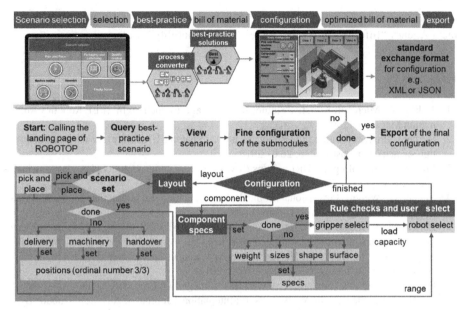

Fig. 3. Exemplary minimum viable configuration process.

5 Conclusion and Future Work

We presented an approach for an architecture concept and a domain cut to build up a microservice architecture for a modular engineering and configuration platform for robotic automation systems. In the context of the research project ROBOTOP we have structured the architecture as a set of loosely coupled microservices. Due to the independency of the various microservices, we can achieve a parallel development, a good extensibility, and a high reusability. The architecture concept described in this paper was developed in project-internal workshops and defines the submodules required for a flexible, modular and scalable configuration platform. The configuration of robot systems is based on the vital best-practice service, which ensures an accelerated design, reduced complexity and modular re-usability.

Future work will focus on the elaborating the architecture based on the operating concept, scientific and psychological recommendations. The operating concept is intended to be a structured approach to divide a complex configuration process into various, interchangeable modules. In addition, methods for the user-specific, skill-based support of the configuration process will be developed.

Acknowledgments

The research and development project ROBOTOP is funded by the Federal Ministry of Economic Affairs and Energy (BMWi). It is part of the technology program "PAICE

Digitale Technologien für die Wirtschaft" and is guided by the DLR Project Management Agency "Information Technologies / Electromobility", Cologne.

References

1. Jackson, J.: Microsoft Robotics Studio: A Technical Introduction. In: IEEE Robotics Automation Magazine (2007).
2. Siemens, Process Simulate for Robotics and Automation, URL: https://www.plm.automation.siemens.com, last visited: 26.01.2018.
3. LogixSim, RoboLogix, URL: https://www.logixsim.com/robologix.php, last visited: 26.01.2018.
4. Michel, O.: Webots: Professional Mobile Robot Simulation. In: International Journal of Advanced Robotic Systems, vol. 1, num. 1 (2004).
5. Bartelt, M., Kuhlenkötter, B. (Hg.): conexing Abschlussbericht. Werkzeug zur interdisziplinären Planung und produktbezogenen virtuellen Optimierung von automatisierten Produktionssystemen. In: Bochumer Universitätsverlag Westdeutscher Universitätsverlag (Maschinenbau, 10). DOI 10.12906/9783899667653 (2016)
6. Ferreira, P., Reyes V., Mestre, J.: A Web-Based Integration Procedure for the Development of Reconfigurable Robotic Work-Cells. In: International Journal of Advanced Robotic Systems, vol. 10, num. 295, pp. 1-9 (2013).
7. Chandrasegaran, S., Ramani, K., Siriam, R., Horvath, I., Bernard, A., Harik, R., Gao, W.: The evolution, challenges, and future of knowledge representation in product design systems. In: Computer Aided Design, vol. 45, pp. 204-228 (2013).
8. igus, Robotik-Baukasten robolink, URL: http://www.igus.de/robolink/roboter, las visited: 26.01.2018.
9. Eynard, B., Lienard, S., Charles, S., Odinot, A.: Web-based Collaborative Engineering Support System: Applications in Mechanical Design and Structural Analysis. In: Concurrent Engineering: Research and Applications, vol. 13, num. 2, pp. 145-153 (2005).
10. Michl, M., Fischer, C., Merhof, J., Franke, J.: Comprehensive Support of Technical Diagnosis by Means of Web Technologies. In: Proceeding of the 7th DET, pp. 73-82 (2011).
11. Fischer, C., Bönig, J., Franke, J., Lusic, M., Hornfeck, R.: Worker information system to support during complex and exhausting assembly of high-voltage harness. In: 5th International Electric Drives Production Conference: Proceedings. Piscataway, NY: IEEE, pp. 212-218 (2015).
12. Fischer, C., Lusic, M., Bönig, J., Hornfeck, R., Franke, J.: Webbasierte Werkerinformationssysteme: Datenaufbereitung und -darstellung für die Werkerführung im Global Cross Enterprise Engineering. In: wt – Werkstatttechnik online, vol. 104, num. 9, pp. 581-585 (2014).
13. Kohl, J., Fleischmann, H., Franke, J.: Intelligent energy profiling for decentralized fault diagnosis of automated production systems. In: Green Factory Bavaria Colloquium (2015).
14. Wolff, E.: Microservices. Grundlagen flexibler Softwarearchitekturen. 1., korrigierter Nachdruck. Heidelberg: dpunkt.verlag (2016).
15. Scholl, B., Swanson, T., Fernandez, D.: Microservices with Docker on Microsoft Azure. Boston: Addison-Wesley (Addison-Wesley Microsoft technology series) (2016).
16. Newman, S.: Microservices. Konzeption und Design. Frechen: mitp. (2015)
17. Torre, C., Singh, D. K., Turecek, V.: Microsoft Azure – Azure Service Fabric and the Microservices Architecture, URL: https://msdn.microsoft.com/en-us/magazine/mt595752.aspx, last visited: 26.01.2018

Ontologies as a solution for current challenges of automation of the pilot phase in the automotive industry

D. Eilenberger [: +49 15222997213] and U. Berger [+49 355-69-2457]

[1] Volkswagen AG, Central Pilot Hall, 38436 Wolfsburg, Germany
daniel.eilenberger@volkswagen.de
[2] Chair of Automation Technology, Brandenburg University of Technology Cottbus - Senftenberg, 03046 Cottbus, Germany
ulrich.berger@b-tu.de

Abstract. The Launch Management is currently playing a more and more important role in the automotive industry. This is mainly due to increasing numbers of variants with decreasing quantities and shortening of the product-life-cycle. A launch can be divided into 2 phases: before start of production (pilot phase) and after start of production (ramp-up). Due to increasing automation in assembly shop and prototype construction, the pilot phase faces new challenges. During the pilot phase the developed facilities and processes must be adopted to fulfill the demands of the series production.
In this paper, the application of ontologies is investigated as a possible solution for the upcoming challenges. For this investigation, at first the characteristics of the pilot phase in the automotive industry are defined. Following the definition, an ontology tailored to the pilot phase's demands is developed. Finally, an outlook will be given towards implementing ontology into future concepts. Additionally an overview of demand for further research will be provided.

Keywords: Pilot phase, Automobile industry, Ontology, industrial facilities.

1 Introduction

One of the most important goals during the pilot phase is to identify and solve problems as soon as possible, targeted at avoidance of rework in the series production and therefore reduction of duration and costs of the launch. Further, main objectives for the launch phase are shortening the entire pilot phase while ensuring product and process quality [1]. Accordingly, the pilot phase represents the base for achieving the three main targets of a launch, which are interrelated. Wildemann describes these targets as reducing the **time for launch**, **securing the quality** of the product and **cutting the costs** of the launch. [2]
The results of Wildemann can be used for weighting the three target values time, quality, and costs. [2] Wildemann has considered various factors and their influence on the accumulated profit of a product. [2] In this context, the negative influences caused by quality deviations and delays during launch are particularly noticeable. Kuhn et al confirms as well, the greatest negative impact on the profit is caused by postponed SOP or

© Springer-Verlag GmbH Deutschland, ein Teil von Springer Nature 2018
T. Schüppstuhl et al. (Hrsg.), *Tagungsband des 3. Kongresses Montage*
Handhabung Industrieroboter, https://doi.org/10.1007/978-3-662-56714-2_19

reduced production volume. [3] Losses caused by a shortfall in production on behalf of low product yields and the shrinking product life cycles in the automotive industry cannot be compensated afterwards. According to this, the shortest possible launch duration and the achievement of the specified quality standards are important success indicators for the pilot phase.

The following chapters gives an overview of the pilot phase's specific characteristics.

2 Characteristics of The Pilot Phase

Some sources indicate the launch phase between the SOP and reaching the targeted volume. [4] Another definition declares the launch phase as a combination of the pre-series, 0-series and ramp-up, with SOP after the 0-series. [5] The later definition is commonly used in the automotive industry, hence that classification will be used in this work (see Figure 1). [6]

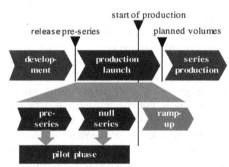

Fig. 1. Pilot phase as a part of the product emergence process (adapted from Schuh, 2008))

Pre-series
During the pre-series, primary production processes and the process technology (such as production facilities and test equipment) are tested. Therefore, prototypes are built by test tools, which are subjected to various product tests in order to evaluate the readiness level for series production of the product and the production processes. At the same time, problems should be identified and corrected at an early stage, as well as, implementation of further process improvements. [1] [7]

0-series
The 0-series is meant for the transfer of equipment and tools into the processes of the series production. From this phase on, series vehicles are manufactured with series tools under series conditions. Meanwhile, internal production processes and logistic processes as well as cross-company processes shared with the suppliers are tested and matched in detail under series conditions. [5] [6] [8]

Ramp-up
Following the SOP, ready-for-sale vehicles are produced for costumers. The production site fulfils the required conditions regarding personnel, organizational, and technical

standards that are necessary to produce the planned vehicle units. These units for the series production are achieved by a successive increase in output quantity until the end of the ramp-up. The aim of the ramp-up is the achievement of the planned volumes under stable production conditions. This target is reached through the implementation of specified quality standards, the scheduled throughput time of the product, the maximum degree of capacity utilization of the production facilities, and the defined costs per unit. [5] [9] [10]

Nagel describes the steady increase of the production volume during the progressing launch, the rise of influencing factors and complexity, as well as the potential damage caused by disturbances. The faults have small effects in pre-series, medium effects in 0-series and large effects in ramp-up. [1]

3 Challenges of the Pilot Phase in the automotive industry

Within the pilot phase, changes to the equipment, the organisation or the process are very expensive due to the complexity of the ramp-up system. This emphasizes the relevance of the pilot phase for a successful completion of the ramp-up project. This relevance becomes clear when considering the three target figures time, costs, and quality of a ramp-up project. [2]

Every pilot phase in the automotive industry is unique. One reason is the increasing number of product variants. The high amount of new technologies that need to be mastered in future launches also leads to the fact that the projects share only few similarities. [4] This effect will be aggravated by the cultural, personnel, and technological differences between the production plants and suppliers of a global automotive manufacturer.

The pilot phase is currently characterized by a high amount of manual labor, due to the demand of flexibility in the manufacturing parameters in the prototype construction. [11] Specially the assembly shop shows very low grade of automation. This fact is changing in the future since it will be possible to also automate assembly processes with new robotic technology. [12] his creates new challenges in the pilot phase, as additional approval for the buildability of parts, the manufacturing equipment and processes must be tested and optimized for the demands of a series production. For handling the pilot phase as effective as possible it is necessary to connect a factory's manufacturing facilities with the manufacturing facilities of prototypes. The experience of already operating facilities can offer great potential for better development of new prototype facilities. [12]

Another challenge in the pilot phase is the increasing digitalization of the factory planning and especially of the prototype construction, also called digital mock-up (DMU). [13] The DMU is a digital model of the new vehicle including its complete structural description. This model simulates all vehicle characteristics required for the construction, planning, manufacturing, or maintenance, and serves as the base for the product and process development. The digitalization of the prototype construction is a component of the so-called digital factory. [14] According to the VDI guideline 4499, the digital factory is an umbrella term for a comprehensive network of digital models, methods, and tools. [15] The digital factory is employed to plan, evaluate, and improve all structures, processes, and resources related to the product. Currently there are several problem areas in this business field. Particularly the lack of data integration, the insufficient heterogenization of the system landscape, and the missing interfaces prevent a complete implementation of the digital factory so far.

Another challenge are the continually compressing product life cycles. [16] [17] Consequently, the number of launches will increase and simultaneously the corresponding pilot phases need to be handled. If there is a vehicle model being produced in different plants at the same time, the amount of pilot phases in the concern multiplies and raises the complexity even more.

According to Opitz and Müller, the increasing technical complexity and fast-paced change in the automotive industry mean that future market developments are increasingly difficult to predict. [4] This, in turn, increases the probability of product design modifications at a very late stage in the launch. Almgren, Terwiesch et al, and Fjällstrom et al point out that these extraordinary product changes can lead to considerable disturbances in the launch phase. [18] [19] [20]

Further challenges of the pilot phase result from its tasks and characteristics within the launch. A essential peculiarity is the great freedom of changing the product or production process. Compared to changes in the series production, these are not so high due to the lower complexity of the launch system and the lower damage potential, but because of their frequency they are a major challenge for the launch management. As described before, the early identification of change requirements regarding the product and the processes is one of the key tasks in the pilot phase. These early changes can prevent problems in the series production or in the material supply and stabilize the launch against disturbances. Identifying and shutting down these change requirements is very complex in a launch project due to the lack of predictability.

4 Ontologies as a solution approach

The increasing automation and digitalization in the automotive industry and especially during the pilot phase offer a variety of options to optimize the interdependent target values launch duration, launch costs, and product quality to each other. Yet the complex challenges of the pilot phase hamper the optimal usage of those possibilities.

Utilizing synergy effects, the time required for developing and testing series equipment and processes can be greatly reduced. A prerequisite for an effective collaboration across the site is the systematic assessment of the complex structures as well as a preferably automated processing of the occurring data. Those structures can be expressed in a multitude of ways, with taxonomy, thesaurus, topic map, and ontology being among the most prevalent models. They are capable of connecting information through relations thus enabling IT systems to process the reality. Differences lie in the respective semantic expressiveness with ontology offering the best solution to simulate complex situations. It is especially convenient for collaborative projects with a larger number of participants, making it ideal to simulate the pilot phase. For such a simulation, an ontology needs to be created that enables the user to identify similar industry equipment and to illustrate the linkage between them within the production chain. The next section describes the essential components of an ontology:

Classes
One of the fundamental components. Classes serve to describe different categories of terms and are usually organized in term hierarchies. This enables the employment of inheritance mechanisms. The topmost class is principally labeled "thing". [21] Each class may manage several subclasses, meanwhile a class may be a subclass of several classes. From the inferior to the superior hierarchy level exists an "is-a relation". [22] [23]

Attributes
Attributes describe properties of classes and objects. The attributes are determined in classes and inherited to the objects. They can be expressed with number, date, and letter values, as well as Boolean values like true or wrong. [24] [23] [25]

Objects
Objects are created through previously defined terms and are the smallest structural units of an ontology. [24] The relations and attributes defined for classes are inherited and are usually also valid for the affiliated instances of the respective classes. [26]

According to Noy and McGuinness there are three fundamental rules which play a substantial role in developing an ontology. Based on those rules and the seven steps of ontology design, the ontology described in chapter 5 was developed. [21]

5   Development of an ontology for pilot phase

In this chapter an ontology especially for the requirements of the pilot phase will be developed.

The first step of the ontology design requires the definition of the ontology domain and scope. The facilities which are expected to be brought to maturity phase during the pilot phase represent the domain. The scope ontology will be determined by the pilot phase relevant areas/divisions. Facilities which are not within the area of responsibility of the pilot phase are outside of the ontology domain or rather the scope.

Next, the important terms for the definite ontology domain pilot phase will be identified. The ontology should reflect the tasks within the scope of the pilot phase as well as the in this context major data production facilities. Essential therefor is the identification and analysis of the relationships between the production facilities within the factory. As already explained does the detail degree depend primary on the questions which should be answered by the ontology based knowledge modelling as well as the of the ontology formulated requirements.

For the definition of the classes will be selected those terms from the previous step, which represent the independent objects. Those terms will therefore defined in classes and structured in upper- and subclasses. The remaining terms which specify those objects will be defined in the following steps of ontology design. The structuring of the terms happens in a combination of the Top-Down- and Bottom-Up approach and with best possible consideration of the guide lines for the number of subclasses. Figure 2 shows the selected and hierarchical structured terms out of step three. The figure merely shows the class structure of the ontology and therefore only an extract of the in this step compiled classes and class hierarchies, as mentioned at the beginning. Illustrated are the fundamental and major hierarchy levels of the ontology. The black framed boxes with three dots indicate that there are further subclasses at this point in the developed ontology but are not shown for a better overview. The upper classes contain predominant two up to twelve sub classes. In exceptional cases, for example in several hierarchic subdevisions of the class "product", an upper class contains more than twelve sub classes. This traces to the high number of vehicle components and the requested detail degree of the ontology.

The upper class „factory area" serves for the detailed identification of a production facility. It consists of the three subclasses „product", „manufacturing methods" and „division". The class "division" represents the all divisions of a car factory in which production facilities have to be brought from pilot phase to maturity phase.

The subclasses shown in figure 2 are based on the structure of a car factory. It is subdivided into „direct divisions" and „indirect divisions". This manufacturing method clustering is based on the DIN 8580 (2003). [27]

The class "manufacturing methods" functions as a detailed classification of the manufacturing facilities and leads to a better comparability.

The class "product" acts as a detailed description of the considered manufacturing facility. The developed structure of the class hierarchy for the upper class "product" reflects the segmentation of a car into different assemblies. It determines which assemblies or parts are manufactured in the according facility. A systematical structuring is the mandatory for the overview due to the high complexity and the vast number of parts of a car. Time consuming searches are countered with adopting established assembly fragmentation. The intuitive application of the ontology provides no necessity for further advices of the cars segmentation. As seen in figure 2 the upper class "product" is divided into the two subclasses "body" and "assembly parts". This segmentation is based on a car's production process. Combined with the class "division" every facility in the factory can be exactly defined. A special focus was given to the maximum flexibility and universality during the development of these classes. This is mandatory for adapting the ontology to the specifics of a launch project like the production site or the car model.

The second upper class "maturity level" is on par with "factory area". This part of the ontology focuses on the manufacturing facility's origin. Following this, focus on facilities from the pre-series, the series, and from tool making can be described. These classes are centralized to connect facilities between pilot phase and series production.

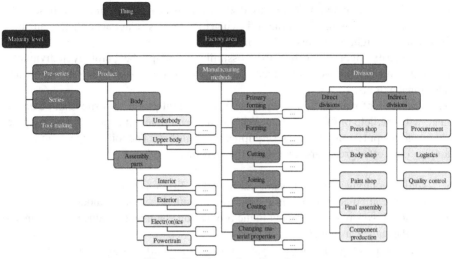

Fig. 2. Ontology for pilot phase

In this paper a fixed definition of attributes is spared to keep the ontology as flexible as possible for further applications.

With the help of the developed ontology it is possible to create object of manufacturing facilities down to single robots. These objects are composed of the selected classes and the filled attributes. Table 1 gives an example of a cockpit assembly facility during the pre-series:

Tab. 1. Example of an Object for cockpit assembly facility during the pre-series

Class		Attribute	
Factory area	-	Location:	Wolfsburg
		Car model:	VW Golf
Product:	Cockpit	Part number:	5N0 881 480
		...	
Manufacturing methods:	Screwing	Torque:	25 Nm
		Screw size:	M6x30
		...	
Division:	Assambley	Robot cell number:	61181
		...	
Maturity level:	Pre-series	Facility number:	PS 61181 03
		...	

6 Possible implementation into a concept and outlook

The ontology created in chapter 5 offers several possibilities for an implementation into a concept for the testing of manufacturing equipment in the pilot phase. A possibility for an implementation consists of extending the existing ontology with additional attributes, relations, or axioms. These extensions are done in a way that the originating ontology is more detailed and better fitted for the desired application.

Following a possible adaption of the ontology the as objects added industrial facilities can be compared with a similarity algorithm. The method described by Juzek can be used for this comparison. [28] The used algorithm is deviated from a case-based reasoning system and is based on the following four factors: class accordance, attribute accordance, weighting factors and axioms.

This similarity algorithm offers the possibility to adapt the ontology individually through adjusting the weighting factor, the accordance of classes and attributes, and additional axioms.

The similarity of manufacturing equipment in different production sites can be calculated with the help of this algorithm. With similar equipment it is possible to compare, share and adapt testing results and equipment parameters. The following set-up for example could offer great synergy effects through sharing of data:

- similar equipment with the same car model in different production sites
- similar equipment for the same standard par with different car models
- up and downstream equipment in the production chain on the same production site for the same car model
- equipment in logistics or quality assurance directly affected by changes
- pre-series equipment of a new car model and similar series equipment of car models already in series production

Another step of implementation into a concept is the adaption of the ontology and the similarity algorithm into a programmed application. Common tools for developing and programming ontologies are OntoEdit, WebOnto, OilEd, or Protégé. Protégé was developed at Stanford University and is provided by free of charge. The application offers an editor for browsing preexisting ontologies and a standard viewer. Protégé is further able to be connect with the application MyCBR which offers the possibility to implement a similarity algorithm.

There is overall still a major demand of research before the created ontology can be implemented into an applicable system. Due to the open and flexible design it is easily possible to adapt to new research results.

7 Conclusion

This paper offers a foundation for solving future issues during the pilot phase of new car models in the car manufacturing business. Challenges originating from increasing automation in assembly shop and prototype construction are especially placed in focus. The characteristics and latest challenges of the pilot phase have been defined for an exact preliminary analysis of the initial situation. This approach is fundamental for creating solutions to reach the goals of the three target values launch duration, launch costs, and product quality.

In this paper the possibility of a systematical analysis of data with an ontology has been used for assessing a possible connection of industrial facilities. For this the basics of creating new ontologies have been described first. Based on this these a tailored ontology has been developed for the demands of the pilot phase. This ontology offers the possibility for creating and comparing objects for every industrial facility of a factory.

An outlook of possible concepts has shown how an exchange of testing results or parameters of similar facilities can work. This paper forms a basis for future tests to achieve a more effective design of the pilot phase in the automotive industry.

References

1. Nagel, J.: Risikoorientiertes Anlaufmanagement. Gabler Verlag, Wiesbaden (2011).
2. Wildemann, H.: Anlaufmanagement: Leitfaden zur Optimierung der Anlaufphase von Produkten, Anlagen und Dienstleistungen. TCW Transfer-Centrum, 13. Aufl., München (2014).
3. Kuhn, A. Wiendahl, H.-P, Eversheim, W. and Schuh, G.: Fast ramp up: Schneller Produktionsanlauf von Serienproduktion, Praxiswissen, Dortmund (2002).
4. Opitz, A., Müller, E. and Hildebrand, T.: Die optimale Anlaufkurve in der Serienfertigung: Ein Modell zur Bestimmung des Aufwand-Nutzen-Verhältnisses von Anpassungsmaßnahmen bei der Beschleunigung des Anlaufs. ZWF Zeitschrift für wirtschaftlichen Fabrikbetrieb, 101. Jg. Heft 6, pp. 356-359 (2006).
5. Laick, T.: Hochlaufmanagement: Sicherer Produktionshochlauf durch zielorientierte Gestaltung und Lenkung des Produktionsprozesssystems, Universität Kaiserslautern. Lehrstuhl für Fertigungstechnik und Betriebsorganisation (2003a).
6. Risse J.: Time-to-Market-Management in der Automobilindustrie. Ein Gestaltungsrahmen für ein logistikorientiertes Anlaufmangement, Technische Universität Berlin. Fakultät VIII Wirtschaft und Management (2002).
7. Laick, T., Warnecke, G. and Aurich, J. C.: Hochlaufmanagement: Sicherer Produktionshochlauf durch zielorientierte Gestaltung und Lenkung des Produktionsprozesssystems. PPS Management - Zeitschrift für Produktion und Logistik, 8. Jg., Heft 2, pp. 51-54 (2003b).
8. Wangenheim, S.: Planung und Steuerung des Serienanlaufs komplexer Produkt-Dargestellt am Beispiel der Automobilindustrie, Peter Lang, Bern (1998).
9. Pfohl, H.-C., Gareis , K.: Die Rolle der Logistik in der Anlaufphase. ZfB Zeitschrift für Betriebswirtschaft, 70. Jg., Heft 11, pp. 1189-1214 (2000).
10. Johnson F. , Karlsson, M.: Ramp-up in manufacturing: Conceptual framework, measurement system and practical experiences from 48 cases in the Swedish industry. Master´s Thesis, Department of Operations Management and Work Organization, Report Number: 98:8, Chalmers University of Technology, Gothenburg (1998).
11. Resch, J.: Kontextorientierte Entwicklung und Absicherung von festen Verbindungen im Produktentstehungsprozess der Automobilindustrie. In: Berichte aus dem INSTITUT FÜR MASCHINEN- UND GERÄTEKONSTRUKTION (IMGK), Band 27, Universitätsverlag Ilmenau, Ilmenau (2016).

12. Weyand, L: Risikoreduzierte Endmontageplanung am Beispiel der Automobilindustrie. Universität des Saarlandes, Saarbrücken (2010).
13. Lawson, G., Salanitri, D., Waterfield, B.: Future directions for the development of virtual reality within an automotive manufacturer. Applied Ergonomics, 53, pp. 323–330 (2016).
14. Bracht, U., Geckler, D. and Wenzel, S: Digitale Fabrik. Methoden und Praxisbeispiele, Springer, Heidelberg (2011).
15. VDI 4499: Digitale Fabrik - Digitaler Fabrikbetrieb (2011).
16. Berger, U., Minhas, S.-u.-H., Lehmann, C. and Städter, J. P.: Reconfigurable Strategies for manufacturing setups to confront mass customization challanges. In: 21st International Conference on Production Research, Stuttgart (2011).
17. Specht D.: Beiträge zur Produktionswirtschaft: Weiterentwicklung der Produktion, Gabler Verlag, Wiesbaden (2009).
18. Almgren, H.: Towards a framework for analyzing efficiency during launch: an empirical investigation of a Swedish auto manufacturer. International Journal of Production Economics, 60/61, pp. 79–86 (1999).
19. Terwiesch, C. and Bohn, R.E.: Learning and Process Improvement during Production Rampup. International Journal of Production Economics, 70, (1), pp. 1–19 (2001).
20. Fjällström, Säfsten, S., K., Harlin, U. and Stahre, J.: Information enabling production ramp-up. Journal of Manufacturing Technology Management, 20, (2), pp. 178–196 (2009).
21. Noy, N. F., McGuinness, D. L.: Ontology Development 101: A Guide to Creating Your First Ontology", Stanford University (2001).
22. Gómez-Pérez, A., Fernández-López, M., Corcho, O.: Ontological Engineering: With Examples from the Areas of Knowledge Management, e-commerce and the Semantic Web. London: Springer Verlag (2004).
23. Dengel, A., Bernadi, A., Elst, L.: Wissensrepräsentation in Semantische Technologien: Grundlagen - Konzepte - Anwendungen, A. Dengel, Ed., Hei-delberg: Spektrum Akademischer Verlag (2012).
24. Haarmann B.: Einführung in die Arbeit mit Ontologien, 2. Aufl. Berlin: epubli. (2013).
25. Bodendorf, Freimut.: Daten- und Wissensmanagement, 2. Aufl. Berlin, Heidelberg: Springer-Verlag (2006).
26. Beißel S.: Ontologiegestütztes Case-Based Reasoning: Entwicklung und Beurteilung semantischer Ähnlichkeitsindikatoren für die Wiederverwendung natürlichsprachlich repräsentierten Projektwissens Dissertation, Universität Duisburg-Essen, Duisburg-Essen (2011).
27. DIN 8580:2003-09 Fertigungsverfahren (2003).
28. Juzek, C.: Entwicklung eines automatisierten, ontologiegestützten Wissensmanagementmodells für Produktionsanläufe in der Automobilindustrie, Shaker Verlag, Stuttgart (2014).

Reconfiguration Assistance for Cyber-Physical Production Systems

André Hengstebeck[1] [0000-0002-9142-2254], André Barthelmey[1] [0000-0002-6111-4592] and Jochen Deuse[1] [0000-0003-4066-4357]

[1] TU Dortmund University, Institute of Production Systems
Leonhard-Euler-Str. 5, 44227 Dortmund, Germany
andre.hengstebeck@ips.tu-dortmund.de

Abstract. In order to overcome today´s challenges of increasing customer requirements, new methods for efficient and customized manufacturing processes have to be developed. This paper presents an approach for the reconfiguration of cyber-physical production systems (CPPS). The concept is based on state-of-the-art virtual plant representations and mapping approaches. First, a suitable virtual plant representation has to be created and overall CPPS capabilities have to be determined. Then the CPPS capabilities need to be mapped with production requirements in order to identify CPPS adaptation needs. In case of necessary reconfigurations, the concept also provides a selection guide in terms of suitable best practices. The concept is applied to a human-robot interactive assembly process. The validation results prove the functionality and high practical relevance of the CPPS reconfiguration assistance.

Keywords: Industry 4.0, CPPS, system reconfiguration, best practices

1 Introduction

Within recent years, the steady increase of international competition initiated major changes regarding the manufacturing industry. Therefore, offering a wide range of individual goods has become a key success factor for manufacturing companies. This often comes along with challenging conditions (e. g. small batch sizes) which may cause high variability within manufacturing systems.

Current developments with respect to Industry 4.0 are highly promising to enable efficient and flexible manufacturing processes [1]. One main focus of this progression is the deployment of cyber-physical production systems (CPPS) that allow for highly flexible adaptation to changing production status in real-time. However, identifying and implementing profound CPPS reconfigurations often requires manual effort by experienced operators. They have to map CPPS abilities with current manufacturing requirements, evaluate the technical feasibility and develop technical solutions.

In this context, the current research projects "PHASE" and "ROBOTOP" study and develop various engineering services for customers, product developers and operators.

© Springer-Verlag GmbH Deutschland, ein Teil von Springer Nature 2018
T. Schüppstuhl et al. (Hrsg.), *Tagungsband des 3. Kongresses Montage Handhabung Industrieroboter*, https://doi.org/10.1007/978-3-662-56714-2_20

Based on production requirements and CPPS capabilities, one of these services provides solutions for individual CPPS reconfigurations and appropriate best practices. As indicated in Fig. 1, the reconfiguration assistance for CPPS consists of three main steps which are a main focus within this paper.

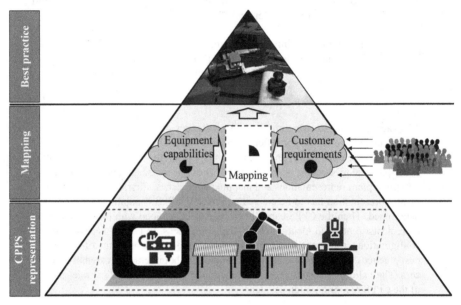

Fig. 1. Concept overview.

2 State of the art

In order to determine necessary requirements and initiate reconfiguration measures, an updated virtual plant representation is necessary. Another key aspect is the mapping of CPPS capabilities with production requirements.

2.1 Virtual plant representation

The virtual representation system is an integral element of a CPPS as it forms a virtual twin of the production system [2–4]. The representation involves data from production, engineering, order processing and other disciplines that can be utilized for various applications and services based on monitoring, control, simulation [5] or prediction [6] along the lifetime of the system.

In the scope of this paper, the virtual representation of machines and devices comprises all relevant information about their components including hierarchy, capabilities and capacities while maintaining completeness and transparency. Therefore, finding suitable technologies which meet these requirements is a major technical challenge. Within international standardization such as ISO/AWI 16400 [7], IEC 62714

[8] and IEC/TR 62541 [9], suitable technologies are currently being discussed in order to become an industrial standard.

The equipment behavior catalogues (EBC) within ISO/AWI 16400 describe capabilities and equipment of machines and serve as a foundation for developing virtual representations. Besides, machine status information can be added to EBCs [10].

The XML based manufacturing data exchange format AML, described within the IEC 62714, is an universal scheme to describe manufacturing devices, machines and plants [11]. AML serves as a single source of information for various use cases throughout the engineering process. The interoperability with the OPC Unified Architecture (UA), a machine to machine communication protocol [9, 12], bridges the gap between engineering and dynamic control data.

In order to meet the requirements of short-cycle adaptions, the virtual plant representation in AML additionally has to include the current system hierarchy. A dynamic plant representation with the objective of automatic plant documentation updates enables feedback from the production plant [13]. Each component serves its own virtual representation whereas slave components communicate their virtual status to the master component. Hence, the current machinery status is continuously updated.

However, the applicability of virtual plant representations for digital production planning and scheduling are still quite limited. Here, the integration of specific planning modules are an important field or research. Especially the digitalization and integration of capability mapping approaches are an important step.

2.2 Capability mapping

Mapping approaches are a central foundation for all kinds of feasibility studies and selection problems. The main idea is to relate specific requirements with performance attributes by means of specific mapping rules. This allows for identifying potential adaptation needs.

Mapping approaches have a long history in terms of skill-oriented automation planning. As early as 1951, an attribute catalogue that included the essential skills of human beings and machines ("Fitts list") was published [14, 15]. Since then, these skill attributes were utilized within different approaches for assessing and comparing skill levels of machines and humans. Well-known substitution-based concepts for function allocation are MABA-MABA charts (Men-Are-Better-At/Machines-Are-Better-At) [16], levels of automation [17, 18] and skill-oriented assembly planning [19–21]. Furthermore, a current concept considers the integration of industrial service robots into existing assembly systems [22].

Beyond the application in automation planning, attribute mapping is relevant to various allocation problems with regard to Industry 4.0. Here, a practical approach focusses on the support of industrial operators with respect to the selection of technical solutions for CPPS [23]. Another current methodology facilitates the selection of smart devices for training of industrial operators on the shop floor [24].

Nevertheless, the automated identification and implementation of CPPS adaptations by means of capability mapping is still a major challenge. In this context, the presented approaches do not allow for real time adaptations. In case of dynamically

180

changing framework conditions (e. g. customer requirements), the whole mapping process needs to be conducted again. Furthermore, the connection of these mapping approaches to ERP / MES systems is still an open point. The benefit of this planning assistance for industrial operators is also quite limited as the main results are often abstract process plans without information on practical use cases. Therefore, a holistic CPPS reconfiguration assistance can create substantial value for industrial operators.

3   CPPS reconfiguration assistance

Based on the current state of the art as well as the identified research gaps, a concept for CPPS reconfiguration assistance was developed. This approach comprises a holistic CPPS communication structure which serves as a main foundation for digital production planning and scheduling. The associated mapping of CPPS capabilities with production requirements allows for dynamic real-time attribute mapping. The identification of potential CPPS reconfigurations is a main decision assistance for industrial operators. In this context, the Best Practice identification supplies users with practical advice regarding the technical implementation of the generated solutions (see Fig. 1).

3.1   Virtual CPPS representation

The input data for identifying potential needs for system reconfigurations has to be extracted from the virtual representation of the CPPS. Therefore, each production resource within the CPPS is permanently equipped with Internet of Thing (IoT) connectors which utilize the existing interfaces for energy supply and field communication (e. g. PROFINET). As the IoT connectors allow for exchanging status data and detecting the CPPS structure (components, hierarchical relations, capabilities), the production resources become cyber-physical systems (CPS).

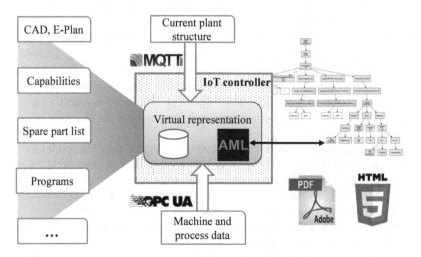

Fig. 2. Exemplary Infrastructure for a virtual representation of CPPS [6].

The overall CPPS communication (see Fig. 2) is event-based (e. g. login of CPS into the CPPS hierarchy), utilizing the IoT Protocol MQTT. Hereby, an overall virtual representation of the CPPS is generated in AML format. This contains the real CPPS structure within the instance hierarchy of AML as well as references to external engineering data (e. g. CAD, PDF, XML). The AML file is converted into a UML data model to enable attribute mapping in the next step.

3.2 Mapping of CPPS capabilities with production requirements

For the mapping process, CPPS capabilities on the one hand and production requirements on the other hand were described by means of attributes. The capability attributes and requirements were structured and connected as a mapping precondition.

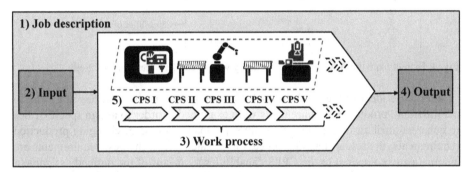

Fig. 3. CPPS model for category definition.

In order to generate a well-structured set of attributes, a CPPS model was compiled on the basis of the work system approach [25]. As indicated within Fig. 3, the CPPS model consists of five major elements which serve as categories for the capability attributes. Here, the job description refers to attributes on the overall work task (e. g. manufacturing technique). The input involves attributes on physical or immaterial manufacturing input (e. g. parts, information). The work processes conducted by the CPS especially involve the executed tasks. The resulting output comprises the manufactured goods as well as additional information (e. g. product quality).

The CPS units are described by attributes concerning their technical features and parameters. Here, the attributes relevant for a specific CPS unit highly depend on its functions. In this context, the Virtual CPPS representation was transferred into an AML instance hierarchy indicated in Fig. 4. This hierarchy helps to identify various functions of CPS units (e. g. robot, gripper, conveyor, control unit, production facility). On this basis, specific sets of relevant capability attributes are assigned to each of these groups. For example, attributes like weight, payload and reach were assigned to the CPS group "robot".

On the other side, the production requirements range from high-level market demands up to specific technical aspects. Therefore, the requirements are divided into the four main categories time, cost, quality and technical feasibility. From the customers' perspective, the first three categories address the main incentives (e. g. increasing number of production units). However, the technical feasibility as a main precondition also influences potential CPPS reconfigurations to a large extent (e. g. required IP code).

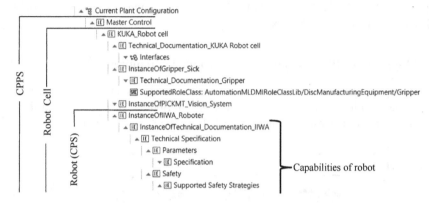

Fig. 4. Extract from the AML instance hierarchy with exemplary plant capabilities.

In order to identify the need for a CPPS reconfiguration, the relations of CPPS capabilities and production requirements were examined in order to assign specific rules to coherent attributes (e. g. object weight < payload). In case of changing production requirements, the actual mapping process is initiated. Here, the new requirements are automatically compared to the CPPS capabilities by means of the defined assignment rules. If the production requirements can be fulfilled by the CPPS capabilities, a positive result is processed. In case of insufficient CPPS capabilities, the concerned attribute and the extent of deficiency are given out to the industrial operator. In order to assist operators in this case, suitable best practices need to be provided as a next step.

3.3   Best practice selection

For the selection and recommendation of appropriate best practices, a database needed to be generated initially. Therefore, best practices were composed by means of literature review and expert interviews with industrial operators. As a main precondition for the selection process, the best practices are structured within a morphology which is based on the presented capability attributes.

In case of insufficient CPPS capabilities, the selection of appropriate best practices is initiated. The underlying selection process is conducted in the same way as described in section 3.2. Thus, only best practices possessing the required capabilities are recommended to industrial operators. If more than one suitable best practice solution is identified, industrial operators can examine all of these to individually derive a reasonable decision for system reconfiguration.

4 Application

The CPPS reconfiguration assistance was applied to a practical use case. According to the presented CPPS model, the job description is to assemble a circulation pump. The input involves five main components (e. g. housing, motor). In regard to the CPS units, there are two lightweight robots, four conveyors with vision systems for material supply, one conveyor for material forwarding, a central control unit and a manual work station. The work process consists of two handling tasks conducted by lightweight robots as well as manual screwing and assembly tasks to produce the output (pump).

For validating the mapping process, the requirement of a new maximum takt time $t_{req}=235s$ per unit is examined as a first step (see Table 1). Within attribute mapping, this requirement is compared with the overall time $\sum t_{process}$ of the manual process. As $\sum t_{process}$ exceeds the required takt time t_{req}, a CPPS reconfiguration is necessary. Therefore, a status report is given to the operator and the best practices selection is initiated.

Table 1. Process time analysis.

Screwing process	Process times					Takt time
	t_{robot1}	t_{robot2}	$t_{screwing}$	$t_{assembly}$	$\sum t_{process}$	t_{req}
Manual	79s	52s	82s	27s	240s	235s
Automated (1)	79s	52s	38 (x2)*	27s	234s	235s
Automated (2)	79s	52s	20s (x4)*	27s	238s	235s

*The manual process involves 8 screws. As the scenarios (1) / (2) only involve 4 / 2 screws, the process times are multiplied.

Here, the unaccomplished production requirements are automatically mapped with the capability attributes of the documented best practices. At this, the database does not comprise a solution that allows for optimizing the robot process times t_{robot1} and t_{robot2}. However, it does contain two best practices for automated screwing ($t_{screwing}$). The automated screwing scenario (1) was analyzed by means of expert interviews at a local company whereas scenario (2) was assessed by literature review [26]. Regarding attribute mapping, only scenario (1) fulfills the assignment rule $\sum t_{process} < t_{req}$ with an overall process time of 234s (see Table 1). Therefore, this scenario is given out as a possible solution to the manual operator (see Fig. 5). If more than one best practice had accomplished the production requirements, all of these would have been given out.

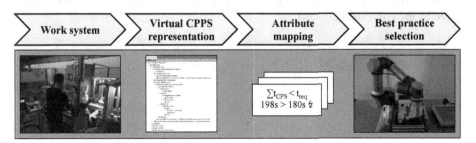

| Work system | Virtual CPPS representation | Attribute mapping | Best practice selection |

$$\sum t_{CPS} < t_{req}$$
198s > 180s

Fig. 5. Overview concept application.

Within the resulting best practice, the Albrecht Jung GmbH & Co. KG utilizes a lightweight robot and a mechanical device for gathering isolated screws, moving them to power adaptors and fastening them with an appropriate gripper. This best practice cannot directly replace the manual screwing process as the framework conditions of these processes differ to a reasonable extent (e. g. components, layout). Nevertheless, the concepts of a mechanical device for screw isolation and an appropriate gripper for the screwing process can be of great use for CPPS reconfiguration. Industrial operators can utilize the given information on the best practices to develop individual automation concepts as well as to evaluate the potential profitability by means of time and invest information. Therefore, the CPPS reconfiguration concept is a valuable decision support for industrial operators within individual planning processes.

5 Conclusion and outlook

Within current industrial practice, the reconfiguration of CPPS is almost exclusively based on experience and expert knowledge, with hardly any help by digital assistance systems. By means of the developed concept, these tasks can be considerably facilitated to generate more efficient planning processes.

As a next step, further CPS groups (e. g. safety devices) need to be analyzed and enriched by capability attributes. Furthermore, additional best practices need to be collected in order to increase the database. Also, the integration of a capacity mapping into the presented approach is important to allow for digital job scheduling assistance. Here, the exchange of capacity data (e. g. machine occupancy) can be supported by the OPC UA interface. For taking full advantage of CPPS, it is crucial to develop a holistic service platform containing the reconfiguration assistance as well as additional modules, such as cross platform simulation, compatibility check or profitability calculation in the future.

Acknowledgement
    The research and development project ROBOTOP is funded under Grant No. 01MA17009H within the scope of the German "PAiCE" technology program run by the Federal Ministry for Economic Affairs and Energy and is managed by the German Aerospace Center in Cologne.
The research and development project PHASE is funded under Grant No. ZF4101110LF7 by the Federal Ministry for Economic Affairs and Energy and is managed by the AiF Projekt GmbH.

References

1. Deuse J, Weisner K, Hengstebeck A, Busch F (2014) Gestaltung von Produktionssystemen im Kontext von Industrie 4.0. In: Botthof A, Hartmann EA (eds) Zukunft der Arbeit in Industrie 4.0. Springer Vieweg, Berlin, Germany, pp 43-49.

2. Barthelmey A, Lenkenhoff K, Schallow J, Lemmerz K, Deuse J, Kuhlenkötter B (2016) Technical Documentation as a Service – An Approach for Integrating Editorial and Engineering Processes of Machinery and Plant Engineers. Procedia CIRP 52:167-172.

3. Uhlemann TH-J, Lehmann C, Steinhilper R (2017) The Digital Twin: Realizing the Cyber-Physical Production System for Industry 4.0. Procedia CIRP 61:335-340.

4. Schuh G, Anderl R, Gausemeier J, Hompel M ten, Wahlster W (eds) (2017) Industrie 4.0 Maturity Index: Managing the Digital Transformation of Companies. acatech STUDIE. Utz, Herbert, München.

5. Boschert S, Rosen R (2016) Digital Twin—The Simulation Aspect. In: Hehenberger P, Bradley D (eds) Mechatronic Futures. Springer International Publishing, Cham, pp 59-74.

6. Strauß, P and Barthelmey, André and Deuse, J (2017) Cyber-physical systems for predictive maintenance. Productivity 2017:12–15.

7. ISO/AWI 16400: Interoperability, integration, and architectures for enterprise systems and automation applications.

8. IEC 62714: Engineering data exchange format for use in industrial automation systems engineering - Automation markup language.

9. IEC TR 62541 IEC TR 62541 OPC unified architecture.

10. Matsuda M, Kimura F (2015) Usage of a digital eco-factory for sustainable manufacturing. CIRP Journal of Manufacturing Science and Technology 9:97-106.

11. Drath R (ed) (2010) Datenaustausch in der Anlagenplanung mit AutomationML: Integration von CAEX, PLCopen XML und COLLADA. VDI-Buch. Springer, Berlin, Heidelberg.

12. Henßen R, Schleipen M (2014) Interoperability between OPC UA and AutomationML. Procedia CIRP 25:297-304.

13. Lenkenhoff K, Barthelmey A, Lemmerz K, Kuhlenkötter B, Deuse J (2016) Communication Architecture for Automatic Plant Documentation Updates. Procedia CIRP 44:365-370.

14. Fitts PM (1951) Human engineering for an effective air navigation and traffic control system. National Research Council, Washington, DC.

15. de Winter, J. C. F., Dodou D (2014) Why the Fitts list has persisted throughout the history of function allocation. Cognition, Technology & Work 16:1-11.

16. Dekker S, Woods DD (2002) MABA-MABA or Abracadabra? Progress on Human-Automation Co-ordination. Congition, Technology & Work 4:240-244.

17. Parasuraman R, Sheridan TB, Wickens CD (2000) A model for Types and Levels of Human Interaction with Automation. IEEE Transactions on Systems, Man and Cybernetics 30:286-297.

18. Miller CA, Parasuraman R (2003) Beyond Levels of Automation: An architecture for more flexible human-automation collaboration. In: Proceedings of the Human Factors and Ergonomics Society 47th Annual Meeting. SAGE, pp 182-186.

19. Beumelburg K (2005) Fähigkeitsorientierte Montageablaufplanung in der direkten Mensch-Roboter-Kooperation. Jost-Jetter, Heimsheim.

20. Bick W (1992) Systematische Planung hybrider Montagesysteme unter besonderer Berücksichtigung der Ermittlung des optimalen Automatisierungsgrades. Springer, München.

21. Westkämper E, Spingler JC, Beumelburg K (2003) Skill Oriented planning of Semi Automated Assembly Systems. In: Proceedings of the 8th IFAC Symposium on Automated Systems Based on Human Skill and Knowledge. Elsevier, pp 111-116.

22. Wantia N, Esen M, Hengstebeck A, Heinze F, Roßmann J, Deuse J, Kuhlenkötter B (2016) Task planning for human robot interactive processes. In: IEEE (ed) Proceedings of the 21st IEEE International Conference on Emerging Technologies and Factory Automation (ETFA).

23. Nöhring F, Wienzek T, Wöstmann R, Deuse J (2016) Industrie 4.0 in nicht F&E-intensiven Unternehmen: Entwicklung einer sozio-technischen Gestaltungs- und Einführungssystematik. Zeitschrift für wirtschaftlichen Fabrikbetrieb 111:376-379.

24. Weisner K, Knittel M, Enderlein H, Wischniewski S, Jaitner T, Kuhlang P, Deuse J (2016) Assistenzsystem zur Individualisierung der Arbeitsgestaltung: Einsatz von Smart Devices zur kontextsensitiven Arbeitsunterstützung. Zeitschrift für wirtschaftlichen Fabrikbetrieb 111:598-601.

25. Landau K, Wimmer R, Luczak H, Mainzer J, Peters H, Winter G (2001) Anforderungen an Montagearbeitsplätze. In: Landau K, Luczak H (eds) Ergonomie und Organisation in der Montage. Hanser, München, pp 1-82.

26. Bauer W, Bender M, Braun M, Rally P, Scholtz O (2016) Leichtbauroboter in der manuellen Montage - einfach einfach anfangen. IRB Mediendienstleistungen, Stuttgart.

Supporting manual assembly through merging live position data and 3D-CAD data using a worker information system

David Meinel[1], Florian Ehler[1], Melanie Lipka[2], Jörg Franke [1]

[1] FAPS, Friedrich-Alexander-University of Erlangen-Nuremberg,
Egerlandstr. 7-9, 91058 Erlangen, Germany
[2] LHFT, Friedrich-Alexander-University of Erlangen-Nuremberg,
Cauerstraße 9, 91058 Erlangen, Germany
David.Meinel@faps.fau.de

Abstract. Most of the existing worker information systems do not base the information on a worker's real movement. By processing live sensor data through a novel Node.js server solution this link was established. This article describes an innovative worker information system with interactive access to real movements and target positions. Especially for assembling safety-relevant components and products, which are bearing a high risk of injury, live-controlled and recorded assembling holds advantages for quality and safety. Moreover, through shortened trainings the efficiency of the whole production system can be increased. The presented worker information system facilitates the employee at the example of a screwing process.

Keywords: Assembly; Information; Worker information system; Radio localization; Smart; Assistance

1 Introduction

Due to new developments of globalization including the growing product variance and the complexity, the industrial production is facing big challenges [1]. In this work, the value of cross-linking and data exchange within manual production gets clearer. In order to address Industry 4.0 and cyber-physical systems this cross-linking should be implemented [1]. The capabilities of machine learning are still restricted and not as applicable as desired. Industry 4.0's goal is thus to integrate the worker into the production in a way that his natural skills, intelligence and creativity can be used most efficiently [2].

The decreasing batch sizes at assembling products and steadily rising costs have a large impact on the current situation for the assembling industry [3]. Therefore, assemblies have to be set up more flexible to provide a better adaptability regarding changes including the type and batch size of products. The integration of the employee into the production process is thus getting more complex and training processes need to be more efficient and faster [4].

© Springer-Verlag GmbH Deutschland, ein Teil von Springer Nature 2018
T. Schüppstuhl et al. (Hrsg.), *Tagungsband des 3. Kongresses Montage*
Handhabung Industrieroboter, https://doi.org/10.1007/978-3-662-56714-2_21

Worker information systems (WIS) are a big support for guiding the employee through production processes. WIS that replace the documentation based training process are often connected to a central database or intranet. In general, such systems are characterized by their purpose and in particular by their field of application [5]. In relation to the way of networking, visualization and degree of mobility, there is a great amount of different WIS types [6].

Real-time monitoring of production processes and interactions with machines and their parameters are only possible by applying real-time systems. This type of software and hardware components is characterized by an asynchronous behavior, meaning which tasks are not executed in a predefined order, but are handled in an event-driven sequence [7]. Two main advantages of real-time systems are the punctuality of response and simultaneity. [8]

Through combining modern localization systems with worker information systems, the tool's real positioning can be verified and corrected. However, most of such existing systems cannot be integrated directly into production systems. Either the accuracy is not good enough or measurement devices are too voluminous to be involved into manual mounting. An appropriate solution could yet be found within the project NaLoSysPro ("Nahfeldlokalisierungssystem Pro") project by using a high frequency localization system which can be mounted remotely [9].

Especially screwdriving represents a great challenge for such localization systems. It requires a high accuracy, precise orientation and small devices [5, 9]. Additionally, the screwing process is one of the commonly used mounting processes in production and can be safety-critical. At the same time, this process is affected by big fluctuations of the screwing results, even when using accurate screwing controllers [10]. Consequently, an efficient and robust controlling of the screwing results is needed to guarantee the best product quality as well as to reduce costs for high-quality products in both later production processes and post processing. Achieving a more efficient and interconnected process to reduce mounting and training time can be identified as another goal of the WIS. Therefore, the connection between real movements and target positions is to be established to save and analyze process data. At the same time, the system should be designed to be intuitive and handy for the worker that any restrictions regarding mobility and mounting sequence can be avoided.

2 Methodology

The goal for the FAPS Institute is to deploy a radio localization system to complex and flexible assembly processes by using the localization data in a WIS for live process control. It is achieved through WIS to guide the user through production processes and to provide information in due course. Therefore, it features an easy and user-friendly operability through its web-based structure. [6]

In order to achieve an elementary information data basis, the system's structure is defined. The implementation of the example screwing process is based on three principles (Fig. 1).

The radio localization system detects the position of the used screwdriver in real-time. It consists of several 24 GHz radar nodes using the FMCW principle [11]. Instead of other localization systems, like ultrasonic or optical systems e.g., it is advantageous for production processes. Ultrasonic systems can be disturbed by engine noises, since they emit waves of similar wavelength. Especially in dusty or very bright environments, conventional optical systems have big problems detecting target positions [12]. A fail-safe operation, which is less influenced by obstacles and the environment, is one of the main goals in radar localization. The used system is composed of three base stations and one transponder mounted onto the screwdriver. While the device featuring the antennas is mostly in control of the signal processing, the transponder can be regarded as a dummy-station. It records signals broadcasted by the antenna device and returns them to the antennas. This outsourcing of complex tasks to a main station allows small and handy transponders [9].

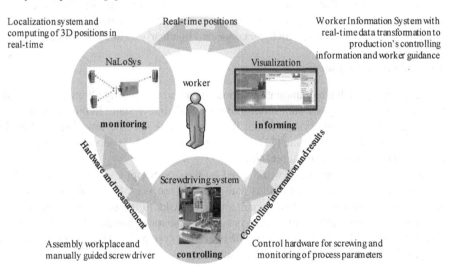

Fig. 1. Principle of an integrated WIS, featuring a screwdriving system, a radio localization system and a visualization

The screwdriving controller is in charge of all required settings for the actual screwing process. It adjusts all parameters according to the task-specific settings provided by the WIS, such as torque, rotational speed and angle of rotation, which are necessary for stable processes. Additionally, the controller features monitoring of the relevant process and environmental parameters. This enables detecting failures during the screwdriving processes, which can be interpreted by the WIS and provide further guiding steps. Informing the user about the current assembly step and the related screwing positions is the main task of the visualization. The integrated live animation of the screwdriver model in the web-based system is issuing a challenge to the researchers of the FAPS Institute.

The animation is based on the position data monitored by the localization system. Through visualizing live data and CAD data, the worker can easily compare his current with the target position. Especially for complex parts with a big amount of varieties, a more efficient mounting process is achieved following this approach.

Since each product requires up to more than one series of working steps, a user is facing a high complexity regarding the involved process knowledge. Every step can contain at least one screwing position. A whole set is summarized to a sequence of screwing positions, of which several can be required. Additionally, one working step can also include safety-relevant messages and assembly notes. Since usually this various pool of information usually demands comprehensive task-specific training, this burden on a user can mostly be eased by the proposed system.

3 Implementation

To guarantee an efficient and robust WIS, despite the high number of different surrounding systems, a consistent implementation is important. Therefore, the system is based on various functions, which are connected by clearly defined interfaces.

3.1 Central data server

Transmitting such a big amount of various data and controlling all interfaces simultaneously needs an asynchronous, non-blocking real-time system. In this context, non-blocking means that the system is not focusing on one specific task, but can handle other requests at the same time [13]. For this reason, a Node.js server is used for handling all data and interfaces in real-time.

Node.js is a real-time framework based on the JavaScript compiler Google V8. It handles requests in an asynchronous and event-driven way. One main advantage of Node.js is the outsourcing of resources-intensive tasks to the operating system. This provides much more efficiency to the system, because it can handle other tasks concurrently during waiting on the answer of the operating system [7]. The implemented data server is founded on six central tasks as following:

- Passing on live data from the nearfield localization system
- Comparison of live data and target positions from the database
- Interface monitoring for error recognition
- Selecting the screwing program and transmitting it to the controller
- Passing on screwing results
- Storing screwing data in the database

3.2 Controlling the screwing process

Due to high fluctuations of the screwing process' results and the related quality costs the result's detection is one of the main task of the WIS. In the case errors are recognized right after mounting, repair costs can be reduced and the product quality can be increased. Considering the screwing process in detail, the most important information to be recorded are:

- transmitted screwing result (OK/failed)

- screwing time (start)
- fastening position (6 DOF)
- recorded position at the start of the screwing process
- product type
- Assigned assembly station

The screwing data is recorded during the whole production process. As soon as the user starts the screwing process, the current position received from the localization system and the point of time are saved. The screwing controller then sends the screwing result via a microcontroller to the Node.js server after the operation. This process will be repeated until all screwing positions of one screwing sequence are finished. Finally, the screwing sequence data is stored into the database.

3.3 Visualization

The visualization is one of the main goals of the implemented WIS, because it provides an augmented reality to the user, in that the next planned steps as well as the current can be overseen and verified. Featuring an animation within the augmented reality (Figure 5, left), safety notes (middle) and information about the screwing process (right), the visualization guides the employee through the production process. The virtual 3D model on the screen copies the movement of screwdriver in real-time, so that the same view like in the real assembly process can be accessed. The screwdriver is positioned in the center of the animation.

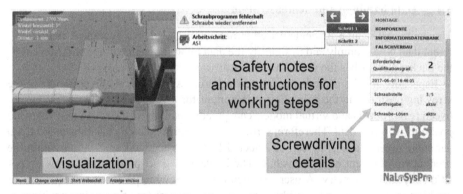

Fig. 2. Visualization of developed WIS with animation and information to current working step

The visualization is based on "Three.js", which is an open-source 3D JavaScript library for web development. It uses WebGL for rendering objects in the web browser [14]. All models are transferrable from real CAD models into the JSON-format and are thus merged into a virtual scene. This whole scene is rendered by the web browser, assembling the product's, the tool's and the environment's 3D representations. To guarantee a realistic view onto the assembly system in its environment, a 360-degree photo is integrated in the scene and is integrated as background through a virtual sphere (Fig. 3, left).

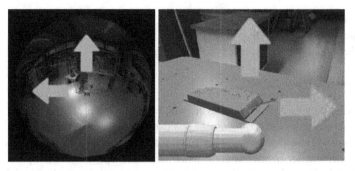

Fig. 3. 3D sphere of an implemented Three.js plant scene (left, the example of the institute's laboratory is used); representation of screwing locations in the 3D model (right).

To guide the user efficiently to the next targeted screwing position, it is visualized by a red arrow, which also indicates the direction for the correct rotation of the screwdriver (Fig. **3**, right). Green Arrows support the process by signaling the necessary movement to get straight to the target screwing position.

When the Node.js server detects valid positioning of the tool, it enables the screwing process by accessing the screwing controller via the Controllino. That is symbolized in the animation by highlighting the virtual screwdriver in green. If the user deviates from the current targeted screwing location, the permission to start screwing process is denied. Additionally, the color of the screwdriver changes back to grey.

3.4 Data handling and demonstration

The main goal of an efficient and well-structured backend is to guarantee, that all different data types comply and to simplify the data input for engineering staff. Moreover, the backend describes the used data in a clear way.

Especially for the demonstrated screwing process, the exact representation of the screwing positions in the animation is important for a valid use. Small deviations between real products and virtual models can already result in wrong relation of screwing positions with live data. Therefore, screwing positions are directly taken from 3D CAD parts and assemblies. They can be selected from an integrated application within the product environment of Siemens NX. The implemented script identifies screwing positions by their geometry. A user can then assign screwing positions to sequences or working steps respectively. Within one sequence, the order of screwing positions is crucial. It determines the order of proceeding through the later steps, which is time-critical and can affect the result's detectability. By subsequently individualizing the order of the screwing positions within a sequence, the user is given the opportunity to address these requirements. Therefore, all data are saved in an Excel file, which contains unique names, coordinates and angles of all screwing locations. The Excel file can then be imported into the database of WIS. Combining the stored data with real time data of the localization system and a clearly defined position of the workpiece (on the assembling table) ensures an exact consistency of all different data.

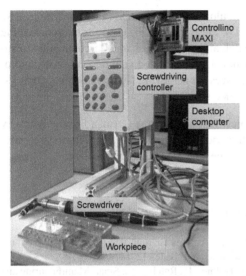

Controllino MAXI	Micro controller for accessing the screw-driving controller from the server
Screwdriving controller	
Desktop computer	Hosts the server and powers WIS
Screwdriver	Tool for applying the screwing operations
Workpiece	Represents actual products

Fig. 4. Example application of assembly systems with a screwdriving controller, a screwdriver, a microcontroller and a product-dummy

To prove the operational capability in real production systems, an example application is built up (Fig. **4**). The screwdriving controller includes 15 different screwing programs, which are chosen from the central data server regarding the current screwing position. In the example application, the system can be tested for different scenarios and on various products. A dummy work piece is used in order to cover a wide range of occurring scenarios, representing an actual product to be assembled.

4   Conclusion and outlook

This paper shows the implementation of a novel WIS that interacts with the user and the environment. Through the combination of a live data visualization in an augmented reality and the connection to the screwing controller as well as to a database, WIS are capable to be more efficient than before. Particularly in the example of guiding the worker through a complex assembly process, manual work can be relived and training reduced.

In the future research, the system's usability as well as its results should be evaluated. Since the localization system was not yet available, as a first approach, a high-accuracy 3 DOF tracking system, such as a laser tracker e.g., will be used as a replacement. Following this approach, the system's functionality can be mostly validated apart from angular derivations of the tool's positioning. Through a tracking system, the WIS could already be validated by evaluating actual screwing processes conducted on the demonstrator. A transfer to other production processes is conceivable, but needs to be examined, at first. Doing so, the transferred data between the tool control and the server has to be adjusted as well as other tool control criteria have to be defined.

In the long run, the process quality should be further increased by analyzing more process data. In the context of industry 4.0, these data can be used for interconnecting different production plants. Future research aims to establish an interconnection to superior control systems. Platform-independent, standardized communication technologies such as OPC-UA e.g., offer great chances for achieving highly interconnected systems and manage to realize sufficient connectivity [15].

Acknowledgements. This research project is funded by the German Federal Ministry of Education and Research (BMBF) within the grant 16ES0160. The authors are grateful for this support and are responsible for the contents of this publication.

References

1. Heinrich, B., Glöckler, M., Linke, P.: Grundlagen Automatisierung. Sensorik, Regelung, Steuerung, 2014th edn. Springer Fachmedien Wiesbaden GmbH, Wiesbaden (2015)
2. Bauernhansl, T.: Industrie 4.0 in Produktion, Automatisierung und Logistik. Anwendung, Technologien und Migration. Springer Vieweg, Wiesbaden (2014)
3. Sanderson, D., Chaplin, J.C., Silva, L.D., Holmes, P., Ratchev, S.: Smart Manufacturing and Reconfigurable Technologies. Towards an Integrated Environment for Evolvable Assembly Systems. In: 2016 IEEE 1st International Workshops on Foundations and Applications of Self* Systems (FAS*W), pp. 263–264 (2016). doi: 10.1109/FAS-W.2016.61
4. Lotter, B., Wiendahl, H.-P.: Montage in der industriellen Produktion. Ein Handbuch für die Praxis, 2nd edn. VDI-Buch. Springer Berlin, Berlin (2012)
5. Fischer, C., Lušić, M., Faltusa, F., Hornfeck, Rüdiger, Franke, Jörg: Enabling live data controlled manual assembly processes by worker information system and nearfield localization system. 5th CIRP Global Web Conference Research and Innovation for Future Production. Science Direct 2017
6. Lušić, M., Fischer, C., Bönig, J., Hornfeck, R., Franke, J.: Worker Information Systems. State of the Art and Guideline for Selection under Consideration of Company Specific Boundary Conditions. Procedia CIRP (2016). doi: 10.1016/j.procir.2015.12.003
7. Cantelon, M., Harter, M., Holowaychuk, T.J., Rajlich, N.: Node.js in action. Manning, Shelter Island (op. 2014)
8. Benra, J.T., Halang, W.A.: Software-Entwicklung für Echtzeitsysteme. Springer, Berlin (2009)
9. Gulden, P., FISCHER, C., Lipka, Melanie, MARSCHALL, A., Fritzsch, T., Dr.-Ing. Franke, J., Dr.-Ing. Vossiek, M.: Nahfeldlokalisierung von Systemen in Produktionslinien. Wireless Localization of Systems in Production and Assembly Lines
10. N. N.: Schraubtechnik und Qualitätssicherung. DEPRAG SCHULZ GMBH u. CO (2014)
11. Brooker, G.: Understanding MillimetreWave FMCW Radars. 1st International Conference on Sensing Technology (2005)
12. Aviles, J.V.M., Prades, R.M.: ARIEL. Advanced radiofrequency indoor environment localization: Smoke conditions positioning. In: 2011 International Conference on Distributed Computing in Sensor Systems and Workshops (DCOSS), pp. 1–8 (2011). doi: 10.1109/DCOSS.2011.5982220
13. Tilkov, S., Vinoski, S.: Node.js. Using JavaScript to Build High-Performance Network Programs. IEEE Internet Computing (2010). doi: 10.1109/MIC.2010.145
14. Dirksen, J.: Three.js essentials. Create and animate beautiful 3D graphics with this fast-paced tutorial. Community experience distilled. Packt Pub, Birmingham, UK (2014)
15. N. N.: OPC Unified Architecture - Specification. Release 1.01 (2009)

Intuitive Assembly Support System Using Augmented Reality

Sebastian Blankemeyer[1], Rolf Wiemann[2], and Annika Raatz[1]

[1] Leibniz Universität Hannover,
Institute of Assembly Technology,
An der Universität 2, 30823 Garbsen
{blankemeyer, raatz}@match.uni-hannover.de,
[2] Leibniz Universität Hannover, student

Abstract. The rising demands of the markets with regard to product individualization and innovative technology deployment force companies to shorten product life cycle times and to use more variants in their production process. To get this flexibility, the employees in the assembly area need to be qualified and supported in an appropriate way. This work presents a concept to support workers during the assembly of products using augmented reality. Head mounted devices are used to demonstrate the correct assembly sequence of the product. The presented data is based on the 3D-CAD models of the product. The different types of parts as well as the positions are detected with the use of optical markers. In that way, changing conditions in the production can be considered. An intuitive interface is implemented to give the worker the opportunity to change program settings tailored to his requirements.

Keywords: augemented reality, assembly instructions, digital workstation

1 Introduction

In a rapidly changing market environment with short product life cycles and a high number of variants, the development of efficient and flexible assembly systems is of particular importance. Assembly is responsible for about 70 percent of the production costs and is therefore of great significance [1]. A complete automation of the processes is often not profitable against the background of fast changing products. Therefore, it is crucial to train employees quickly in assembly processes in order to achieve a short start-up time.

Augmented Reality (AR) offers new perspectives in this respect. In contrast to paper- or video-based instructions, AR makes it possible to visualize real-time information directly in the real environment. The effectiveness of AR systems for supporting and training in assembly processes has already been proven. Thus, M. Funk [2] could show that AR-guidance could help to execute a assembly task faster and with fewer errors compared to a video-guidance. Blattgerste et al. [3] investigated the suitability of different AR-devices (Smartphone, Microsoft

© Springer-Verlag GmbH Deutschland, ein Teil von Springer Nature 2018
T. Schüppstuhl et al. (Hrsg.), *Tagungsband des 3. Kongresses Montage*
Handhabung Industrieroboter, https://doi.org/10.1007/978-3-662-56714-2_22

Hololens, Epson-Moverio). They came to the conclusion that the error rate could be reduced by using the Hololens. With the Epson-Moverio, a slightly shorter execution time compared to the Hololens was measured.

The Microsoft Hololens connects an AR system with a camera system in a compact device. In contrast to other data glasses (e. g. Google Glass, Epson-Moverio) the environment can be tracked spatially and virtual content can be visualized three-dimensionally. In addition, it is possible to control the Hololens and interact with holograms via gesture and language. For this reason, in this work a Hololens is used as an appropriate tool to train and support employees by simulating an assembly process directly at the workplace.

In this article, a short overview of existing assembly applications using augmented reality is given. After that the concept of the support system as well as the implemented use case are described. Based on the conclusion, ideas for future work are presented.

2 Augmented Reality in Assembly

In general, each AR system consists of a camera unit for tracking the environment and a display system for displaying the virtual content. The technology can be divided in three different applications: Head-mounted displays (HMD), handheld devices and projection-based systems. Various input devices such as data gloves, markers or gesture and speech recognition are used to control the AR application and to interact with the holograms. Through the development of high-performance AR hardware and software in the recent years, there are already some applications of AR systems in gaming and production industry. However, research is still the largest field.

AR-systems are used in the industrial environment mainly in the field of assembly and maintenance [4], [5]. In this way, employees could be supported with additional real-time informations of the process. For example, Boeing uses Google Glass to visualize assembly instructions in the field of wiring harness manufacturing and has been able to reduce production time by up to 25 percent [6].

One area of research is the use of projection systems for the visualization of assembly steps and instructions. Rodriguez et al. [7] used a projector to visualize assembly instruction on the workplace. With a recognition system the workflow of the user is observed to display the information dynamically. M. Funk et al. [8] developed several systems within the context of the research project "motionEAP". The systems support the user during assembly and commissioning tasks. Information about the assembly progress and assembly sequence can be visualized at the workstation using in-situ projection. A Microsoft Kinect detects the actions of the user in order to display information about the current and the next assembly process. Another system, which is also based on a projector, is used to provide picking information.

A similar system for the visualization of assembly information was developed by O. Sand et al. [9]. Picking information and assembly instructions can be displayed using a projector at a workstation. This system was also implemented with an HMD applying an optical see-through display [10] to the employee. The picking containers can be detected by optical features and thus without the need of markers. Buttner et al. [11] compared the HMD-system with the projector based system and came to the conclusion that a projection system is better suited for their usage. The main problem for a widerange industrial usage of HMD are the wearing comfort and display quality.

G. Michalos et al. [12] uses AR in the field of collaborative assembly. Assembly steps are visualized on a tablet and information about the robot status is displayed. X. Wang et al. [13] realized a system based on an HMD and a stereo camera for object and also gesture recognition. Using this system it is possible to simulate virtual assembly steps on the one hand and to perform the assembly virtually with gestures on the other hand.

Unlike most other HMDs, Microsoft Hololens not only allows to place holograms in the environment in a three-dimensional and realistic way but also enables interaction with the holograms via gestures. R. Radkowski et al. [14] use the Hololens to visualize assembly steps. With an additional Microsoft Kinect, objects are detected and can be virtually overlapped with the reality. G. Evans et al. [15] developed a Hololens application to display assembly information as a textual instruction. The individual components and groups are detected via marker tracking.

In current applications mainly projection systems are used for assembly support. The advantage of these is that the user does not have to carry an additional device. However, the disadvantage is that the system is permanently installed and therefore inflexible. HMDs allow variable usage, but in some cases they have poor display quality, poor wearing comfort and no three-dimensional representation. Nevertheless, the Microsoft Hololens offers the advantage of a spatial representation and a good display quality. The suitability of the Hololens has already been investigated and proven by Blattgerste et al. [3] and Radkowski [14]. Despite the small field of view and the wearing comfort for longterm use, the Hololens is a very powerful device and offers a wide range of possibilities. In addition, the Hololens supports 3D objects which are more intuitive for the workers than 2D pictures of the process. Taking this into account, in this application the advantages of HMD like the Hololens are used to give employees an intuitive tool to support their work.

3 Intuitive Assembly Support System

The goal of this work is to develop a Hololens application to visualize assembly steps of a product. The programming and handling should be as flexible as possible. On the one hand, it should be possible to change the product without major programming effort. On the other hand the application should be usable in various environments (see Fig. 1).

To achieve this goal, the worker is wearing the Hololens so that he could use both hands for the assembly process. Since the Hololens is a see-through device

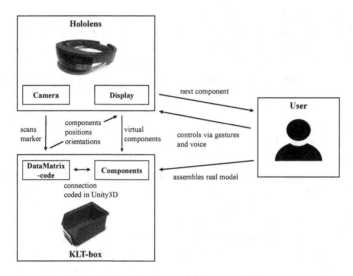

Fig. 1: Concept of Support System

the worker is able to see virtual objects implemented in his environment. This feature can be used to display the needed object for the next assembly step. To support the employee, the start and the end position as well as the path between them are shown in the device. After assembling one component, the worker can choose the next object. In addition, a mix up of components is prevented by using this technology.

In order to design such an assembly instruction for every product a 3D CAD-model is mandatory. This CAD model has to be divided into its individual components. Optical markers are assigned to the components in order to detect the positions in the real environment and to overlap the real components with the virtual ones. The real model has to be assembled on a mounting aid or a defined assembly area, which position is detected via marker tracking as well. The AR software developer Vuforia offers a *Software Development Kit* (SDK) for the Hololens, which is used for the marker tracking. Conducting this step, the virtual CAD-objects are connected with the given environment at the working station, what is essential to determine the start and end position.

The interaction with the virtual objects is realized via the Hololens' standardized communication interfaces. The direction of view controls a cursor and the user could interact with the focused object via an AirTap, comparable to a mouse click. Based on this, manipulation gestures can be used to move the virtual model. Speech recognition and output is also possible. The developed Hololens application can be controlled via buttons as well as voice commands.

To make the application more flexible, the user should be able to determine and change the sequence of assembly. A menu is implemented for this purpose, with which the order of the components can be defined by a simple drag-and-drop principle. The sequence can be changed if the predefined assembly sequence proves to be impractical. As a further function, the entire demonstrator model can be displayed. This gives the user an overview of the model. It can also be shown during the actual assembly process to recall the sequence.

4 Implementation and Use Case

Microsoft Hololens uses the Windows 10 operating system and the *Universal Windows platform* (UWP) as the unified development platform for Windows 10 applications. For the development of Hololens applications, Visual Studio is used in combination with the game development software Unity3D, for which Microsoft already provides fundamental elements such as gesture recognition.

The current application is implemented and tested at the working stations of the learning factory at the Institute of Production Systems and Logistics (IFA) at the Leibniz Universität Hannover. The subject of the assembly is a helicopter model (see Fig. 2), which is mainly used as a demonstrator in the given learning environment. To display the model holographically, the CAD model is first converted into a mesh model to be usable in Unity3D. This step has to be done for every new CAD model once again. The program logic created for this application can be applied to any other model in Unity3D. The markers and program scripts only have to be assigned to the components, when programming a new product or variant. This enables an easy change of the product.

Fig. 2: Assembled helicopter (left) and DataMatrixCode with KLT box (right)

The real parts are provided in KLT boxes, which are equipped with an electrophoretic display, that are usually used to state out the prices in grocery stores (see Fig. 2). The content of the displays can be configured via a software running on a server system. This server can be connected to the production system to get the correct datacode corresponding with the content of the box. In this way, failures in the assembly process can be reduced, since a mix up of components can be prevented due to the fact that the assembly instructions in the HMD are connected to the KLT boxes via the optical markers.

In order to simplify the usage and to respond the worker's demands two interaction methods are implemented to control the application. The main interface is a toolbar with buttons to configure all settings. The toolbar can be placed everywhere in the near of the working station where it does not impair the worker.

In addition, a voice recognition is implemented to get a more intuitive way of interaction with the Hololens.

4.1 Marker-Tracking

Markers are used to detect the positions of the virtual modules in the KLT boxes and the mounting aid on the worktop. With the Vuforia SDK 2D-markers of any design can be detected. The markers used in this paper are based on 2D codes of the DataMatrix-code type. These codes often do not fulfil all the requirements of Vuforia, which are recommended for the most accurate tracking. Nevertheless, they are used in this work because they are well suited for the electrophoretic display at the KLT boxes and the tracking accuracy is not crucial for the application (see Fig. 3). The markers for the modules are shown on the

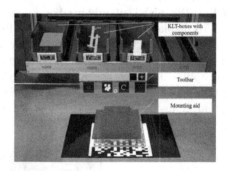

Fig. 3: Assembly station with scanned markers

displays on the KLT boxes. The marker of the mounting aid can be placed anywhere on the work surface. The position of the virtual parts to the markers is defined in Unity3D. As soon as the markers have been detected, the virtual components are visualized. The Vuforia tracking is then switched off to save computing resources. However, it can be switched on again at any time to rescan the markers if the situation has changed.

4.2 Simulated Assembly

To realize a virtual assembly of a product, the model has to be split into the individual components. In this paper, the application is implemented for three components but it can be extended easily. The assembly structure and assembly sequence is transferred from the CAD model and can be adapted in Unity3D. In addition, the user of the application can change the order of the sequence tailored to his individual requirements. For this purpose, a list is implemented in which all components are displayed in their assembly order. The order can be rearranged by moving the list elements with a manipulative gesture (see fig 4a).

Each component is assigned to one marker. After the application has been started, all markers need to be scanned with the hololens. The start position is defined by the scanned markers of the KLT boxes. The target position of each assembly is defined on the mounting aid through the CAD model of the whole assembly. When the assembly process/instruction is started, a linear path is interpolated between the start and end position on which the assembly is moved. In addition, a rotation is interpolated with the spherical quaternion interpolation (Slerp) between the start and target orientation. One assembly step is simulated as long as the next one is called. For a better overview, only the currently selected object is displayed. Once all components have been placed, the program could be reset and the simulation can start again.

In order to additionally support the employee during his work the entire helicopter model can be shown. Using this opportunity, the worker get a better overview of the current assembly step in relation to the whole product. The model can be placed, rotated and scaled with a manipulative gesture at the working station (see Fig. 4b). Although the communication interfaces of the

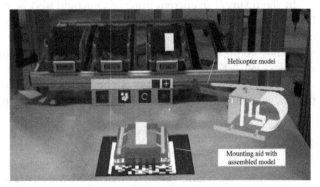

(a) Assembly sequence

(b) Workstation with virtual components and entire helicopter model

Fig. 4: Assembly workstation

Hololens enable intuitive control of the program, there is still a limitation of the HMD concerning the small field of view. If the user is too close to the workplace, not all holograms are visible at the same time. Nevertheless, the display quality is convincing. Regardless of the brightness of the environment, holograms can be easily recognized.

The approach offers the worker a powerful assembly support system. If CAD models of the components exist, the application can be easily implemented. The advantage for the user is that the assembly sequence is presented in an intuitive form and both hands can be used to conduct the assembly process.

5 Conclusion and Outlook

In this paper, a concept for an assembly support system based on augmented reality was presented. Using a head mounted device, the worker gets information about the next assembly sequence as 3D-holograms to simplify the assembly process of whole models. The sequence can be changed by the user as required. Additionally, the entire model can be displayed to get an overview of the model. The application itself can be used flexibly and is not limited to a special workstation. Only the markers for the components need to be scanned to connect the reality with the virtual components. An intuitive interface based on gesture and speech recognition is implemented to set up the program.

In order to be able to use the application even more flexibly, further research deals with the topic that no programming with Unity is needed anymore to change the model. This would be possible by moving a large part of the programming in Unity3D to the Hololens application. For this, an interface to a cloud such as *OneDrive* or *Dropbox* could be implemented, through which both a model and the corresponding markers can be loaded. Markers are then assigned to the individual components to set up the assembly sequence.

In the future, the current application could also be expanded into the area of human-robot collaboration. Robot programming often requires specialist knowledge and is therefore an obstacle to use robots in production systems. Programming could be simplified with the intuitive communication interfaces of the Hololens. With the current application, the positions of the components can already be determined via markers. Coordinates could then be exchanged between the Hololens and the robot via a defined, common coordinate system.

References

1. Lotter, B., Wiendahl, H.P.: Montage in der industriellen Produktion (2012)
2. Funk, M.: Augmented Reality at the Workplace : A context-aware assistive system using in-situ projection. Ph.D. thesis (2016)
3. Blattgerste, J., Strenge, B., Renner, P., Pfeiffer, T., Essig, K.: Comparing conventional and augmented reality instructions for manual assembly tasks. Proceedings of the 10th International Conference on PErvasive Technologies Related to Assistive Environments - PETRA '17 pp. 75–82 (2017). DOI 10.1145/3056540.3056547
4. Fraunhofer-Anwendungszentrum Industrial Automation: smartfactory-designspace (2017). URL http://www.smartfactory-owl.de/designspace/designspace.html
5. Evaluating the application of augmented reality devices in manufacturing from a process point of view: An AHP based model. Expert Systems with Applications **63**, 187–197 (2016). DOI 10.1016/j.eswa.2016.07.006
6. Sacco, A.: Google glass takes flight at boeing (2017). URL http://www.cio.com/article/3095132/wearable-technology/google-glass-takes-flight-at-boeing.html
7. Rodriguez, L., Quint, F., Gorecky, D., Romero, D., Siller, H.R.: Developing a Mixed Reality Assistance System Based on Projection Mapping Technology for Manual Operations at Assembly Workstations. Procedia Computer Science **75**(November), 327–333 (2015). DOI 10.1016/j.procs.2015.12.254

8. Funk, M., Kosch, T., Kettner, R., Korn, O., Schmidt, A.: motionEAP: An Overview of 4 Years of Combining Industrial Assembly with Augmented Reality for Industry 4.0. Proceedings of the 16th International Conference on Knowledge Technologies and Data-driven Business pp. 2–5 (2016)

9. Sand, O., Sebastian, B., Paelke, V., Carsten, R.: Virtual, Augmented and Mixed Reality **9740**, 643–652 (2016). DOI 10.1007/978-3-319-39907-2

10. Paelke, V.: Augmented reality in the smart factory: Supporting workers in an industry 4.0. environment. 19th IEEE International Conference on Emerging Technologies and Factory Automation, ETFA 2014 (2014)

11. Büttner, S., Funk, M., Sand, O., Röcker, C.: Using Head-Mounted Displays and In-Situ Projection for Assistive Systems. Proceedings of the 9th ACM International Conference on PErvasive Technologies Related to Assistive Environments - PETRA '16 pp. 1–8 (2016). DOI 10.1145/2910674.2910679

12. Michalos, G., Karagiannis, P., Makris, S., Tokçalar, Ö., Chryssolouris, G.: Augmented Reality (AR) Applications for Supporting Human-robot Interactive Cooperation. Procedia CIRP **41**, 370–375 (2016). DOI 10.1016/j.procir.2015.12.005

13. Wang, X., Ong, S.K., Nee, A.Y.: Multi-modal augmented-reality assembly guidance based on bare-hand interface. Advanced Engineering Informatics **30**(3), 406–421 (2016). DOI 10.1016/j.aei.2016.05.004

14. Radkowski, R., Ingebrand, J.: Virtual, Augmented and Mixed Reality **10280**, 274–282 (2017). DOI 10.1007/978-3-319-57987-0

15. Evans, G., Miller, J., Iglesias Pena, M., MacAllister, A., Winer, E.: Evaluating the Microsoft HoloLens through an augmented reality assembly application **10197**, 101,970V (2017)

GroundSim: Animating Human Agents for Validated Workspace Monitoring

Kim Wölfel, Tobias Werner, and Dominik Henrich

Chair for Robotics and Embedded Systems,
Universität Bayreuth, D-95440 Bayreuth, Germany,
kim.woelfel@uni-bayreuth.de,
http://robotics.uni-bayreuth.de/

Abstract. In the promising field of human-robot cooperation, robot manipulators must account for humans in the shared workspace. To this end, current prototypes integrate various algorithms (e.g. path planning or computer vision) into complex solutions for workspace monitoring. The step from research to industrial use for these solutions demands rigid validation of the underlying software with real-world and synthetic data. Related fields (e.g. human factors and ergonomics) implement toolsets to create synthetic data of human-machine interactions. However, existing toolsets employ hand-crafted motion paths or motion segments for their human agents. This limits the variety of resulting motions and implies laborious composition of animation sequences. In contrast to this, we contribute a novel approach to human animation for synthetic validation: We animate our human agents through a realistic physics simulation and we expose motion paths in a flexible and intuitive high-level editing interface. We also generate photo-realistic images of resulting animations through state-of-the-art rendering techniques. Finally, we employ these synthetic images and their ground-truth backing to validate a prototype for a workspace monitoring system and a subsequent online path planner.

Keywords: workspace monitoring, virtual environments, ground-truth testing, synthetic data, physically-based animation of humans, computer vision, robotics

1 Introduction

Traditional industrial manipulators execute automated tasks in isolated robot cells. However, the recently emerging field of robot-human cooperation envisions a shared workspace for robots and humans. Robots, particularly, must perceive human agents and other a-priori unknown objects in the robot cell in real-time to determine appropriate on-line reactions. Prototypical monitoring solutions for the shared human-robot workspace (e.g. [23]) suggest algorithms of computer vision to derive workspace occupation through varied sensor data. Subsequent stages for collision checking and path planning can incorporate the reconstructed workspace occupation to find a suitable reaction for the robot manipulator. Example reactions include speed control or adjusting the path of the robot.

© Springer-Verlag GmbH Deutschland, ein Teil von Springer Nature 2018
T. Schüppstuhl et al. (Hrsg.), *Tagungsband des 3. Kongresses Montage Handhabung Industrieroboter*, https://doi.org/10.1007/978-3-662-56714-2_23

Both monitoring systems and robot reactions must be validated thoroughly for the step from research to real-world applications. Full validation obviously requires real-world data. However, real-world data is not without its drawbacks. Acquiring extensive real-world data is time-consuming and error-prone, the robot cell must be available and remains occupied during tests, dummy workers and example obstacles are required, and finally no ground-truth data exists for test automation (e.g. for unit and integration tests).

Synthetic testing data promises to solve the above problems of real-world data at the cost of fidelity. The significance of validation increases with the fidelity of synthetic testing data. To test system limits, fidelity is particularly relevant for traditional failing points of computer vision: On the one hand, texturing and lighting of the rendered scenes must be photo-realistic for meaningful validation of background subtraction and segmentation. On the other hand, movement of human agents must be lifelike for meaningful validation of time-coherent obstacle detection (e.g. for approaches with tracking or online learning).

Traditional solutions use off-the-shelf rendering and animation techniques such as basic Phong lighting and animations stitched together from motion-captured segments. Opposed to this, our contribution excels in two points: We combine state-of-art real-time rendering (e.g. shadow mapping, normal mapping) with a physics simulation to drive realistic human movement. Additionally, our toolset offers a high-level and intuitive way to specify flexible motions for human workers without relying on rigid motion captures or laborous motion stitching.

2   Related Work

Over recent years, robotics simulators have become widespread. Such simulators enable the user to design and validate robot applications in virtual environments, including simulated sensors such as depth cameras or laser scanners. Some of the simulators (e.g. RoboDK [18]) focus on the simulation of robotic manipulators, partially even without a graphical user interface (GUI) (e.g. [11]). Other simulators try to cover a wide variety of robot types (e.g. Actin [1], Gazebo [10], UARSim [3], OpenRAVE [5], V-Rep [7], Webots [16]).

Closest to our contribution are simulators that offer a specific set of features: simulated physics with collision detection, photo-realistic rendering, animated human agents, a graphical user interface, and virtual sensors. Consequently, we chose Gazebo, V-Rep and UARSim for an in-depth review.

The Gazebo framework is a comprehensive simulator which is well established in the scientific community. Gazebo supports multiple high-performance physics engines (e.g. ODE [14] and Bullet [2]), it utilizes the Open source 3D GRaphics Engine (OGRE [15]) for realistic rendering and it includes virtual sensor models with or without simulated noise. Finally, Gazebo animates human agents based on predefined joint trajectories. Our contribution, in contrast, generates human movements at runtime based on a physics simulation.

The V-Rep framework is another toolset for validating robotics applications. V-Rep offers fast collision detection and a choice of four physics engines (Bullet,

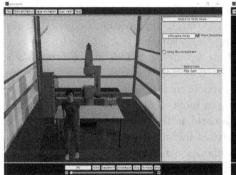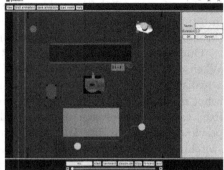

Fig. 1: GroundSim toolset: Sensor View (left) and Edit View (right)

ODE, Newton [13] and Vortex Dynamics [20]) to simulate real world rigid-body physics. The integrated rendering engine performs photo-realistic rendering with soft shadows. Virtual sensor models and a graphical user interface round out the features of V-Rep. Opposed to our contribution, V-Rep animates human agents by combining predefined motions.

The UARSim framework is a lightweight robotics simulator for research and education. UARSim builds upon the Unreal Engine, it offers virtual sensors, and it comes with a graphical user interface. Finally, UARSim can simulate human agents, but does so by means of predefined motion trajectories.

3 System Overview

This section describes the main aspects of our toolset. We then continue to highlight features of the photo-realistic rendering component, we discuss advantages of working with synthetic data, and we explain our approach to the physics-based animation of human agents.

3.1 Overview and Main Features

The GroundSim toolset serves as a **Sim**ulator for generating **Ground** Truth Data. It enables rapid validation of robotics applications. We are particularly concerned with applications that perform online monitoring of shared workspaces for adaptive path planning. To this end, GroundSim allows users to intuitively and effortlessly generate animation sequences of virtual human workers inside a robot cell. Photo-realistic rendering of these animation sequences subsequently produces virtual sensor input (e.g. as a simulated multi-camera system). Finally, combining virtual sensor input with available ground-truth backing data supports the thorough validation of software components, including online monitoring and path planning.

Our approach differs from other toolsets in how users choose the motions of human workers in the robot cell. We do not use predefined motion trajectories, but calculate motions on the fly with a physics simulation. Our graphical user interface (see Fig. 1 (left)) enables users to specify motion sequences for workers in a robot cell through worker positions, through per-position behaviours, and through behavioural segments connecting the positions.

Our toolset operates in two different modes: *Sensor Mode* and *Editing Mode*. When operating in sensor mode (see Fig. 2 (left)), the center viewport shows a realistic rendering of the current workspace, including human agents and the robot manipulator. While arbitrary camera movement is possible, we also have fixed camera presets that mimic a real-world multi-camera setup for intuitive preview of later output. Additional features include changing object visibility, switching between photo-realistic and silhouette rendering modes (see Fig. 2), and toggling the display of obstacle bounding boxes.

In editing mode, the virtual camera of the center viewport is fixed to a bird's-eye view (see Fig. 1 (right)). Users are then able to define an animation path for the human worker by using two kind of animation primitives, *animation nodes* and *animation segments*. In particular, users place animation nodes both at positions where the human worker should interact with the environment and at goal positions for walking animations. Our toolset automatically links all consecutive animation nodes with animation segments. The planning algorithm for animation segments divides the floor into small, square tiles, thereafter adds the scene objects as obstacles, and finally executes the A* algorithm [9] to find a collision-free path for the human worker.

The resulting path p consists of a set N of animation nodes n_i, connected by a set S of animation segments s_j (see Fig. 1 (right)), where the count of animation segments c_s depends on the count of animation nodes c_n, i.e. $c_n = c_s + 1$. Users can assign a set of behaviours B_{node} or B_{segment} to each primitive of p, including behaviours such as walking, waving a hand, picking up an object, or shoving an object with the foot.

For animation nodes, we offer reasonable behaviours,

$$B_{\text{node}} \subset \{b_{\text{stand}}\} \times \{\emptyset, b_{\text{wave\_hand}}, b_{\text{kick}}, b_{\text{pick\_up}}, b_{\text{drop}}, ...\}, \qquad (1)$$

while for animation segments, we expose different behaviours,

$$B_{\text{segment}} \subset \{b_{\text{walk}}, b_{\text{run}}, ...\} \times \{\emptyset, b_{\text{wave\_hand}}, b_{\text{drop}}, ...\}. \qquad (2)$$

Individual behaviours are physically grounded due to the use of a physics-based animation kernel (see [4]). When generating the animation, we merge individual behaviours of each animation primitive to an overall action. This may, for instance, result in an animation sequence that combines walking with waving the hands.

3.2 Rendering Engine

For the photo-realistic visualization of animation sequences, we apply a custom rendering engine. Our engine [21] follows a command-driven architecture, where C++11 shared pointers manage opaque IDs over data of naive rendering APIs. Notably, client applications such as our toolset can issue rendering commands

Fig. 2: Photo-realistic (left) and silhouette (right) rendering modes.

(e.g. to update graphics resources or to compose virtual scenes) to data IDs from arbitrary client threads. This enables convenient integration of additional software components such as monitoring systems and path planners for synthetic, but still online validation. Supported back-ends include a photo-realistic OpenGL 3.2 path, a fallback OpenGL 1.1 path, and an NVIDIA Optix GPU ray-tracing path. The OpenGL 3.2 path features state-of-art rendering techniques such as deferred lighting (see [8]), shader-based normal mapping (see [12]), and smooth shadows through shadow mapping (see [6]). Opposed to game-centric engines (e.g. OGRE), our rendering kernel efficiently can exchange data with other software components. For example, rendering to CUDA-capable offscreen back-buffers enables overhead-less coupling with fast GPU implementations of computer vision algorithms. Finally, we use a custom GUI module to visualize arbitrary 3D viewports, for example to realize a preview of all virtual sensors.

3.3 Advantages

The advantages of our contribution for the validation of robotics applications are twofold: We supply realistic, but synthetic data instead of real-world data and we simulate characters based on a physics simulation. With respect to application validation, synthetic data surpasses its real-world equivalent in notable cases. We discuss select cases in the following, alongside respective advantages of our toolset. See Table 1 for a summary of our discussion.

We first consider validating the *workspace occupation* as an output of the monitoring system. Depending on the later application, monitoring systems build subsymbolic (e.g. voxels) or symbolic (e.g. transforms, poses) representations of workspace occupation. In real-world validation ground-truth data is not available implicitly for either kind of representation. Existing approaches thus measure or derive approximate ground-truth by limited, laborous, invasive, or expensive procedures, such as laser scanning or high-fidelity motion capturing. Opposed to this, synthetic data as generated by our toolset comes with an implicit, effort-less, inexpensive and exact ground-truth: Positions, orientations and geometric meshes of obstacles and humans are readily available. Finally, we can opt to

Table 1: Comparison of real-world and synthetic data

Trait	Real-World Data	Synthetic Data
Workspace Occupation	not exact	ground-truth
Segmentation	not exact	ground-truth
Sensor Calibration	time consuming	fast and optional
Workcell Access	necessary	not necessary
Time	fixed	speed-up possible
Coverage	manual	permutations and fuzzying
Repeatability	hardly possible	possible
Modification	hardly possible	possible
Hazardous Scenarios	no	possible

perform *camera calibration*, or we can use the implicitly known, ground-truth camera parameters to correlate workspace occupation with later path planning.

Foreground-background *segmentation* [17] constitutes a rather sensitive part of monitoring systems. Therefore, the individual validation of respective software components is worthwhile. Since there is no exact ground-truth for real-world workspace occupation, the standalone validation of segmentation proves difficult. Options include hand-labeling of chosen images with an optional pass through a slow, but high-precision classifier (e.g. AdaBoost on select image features) for the remaining images. Our contribution, however, can directly generate exact ground-truth segmentations from ground-truth workspace occupation.

Apart from ground-truth concerns, synthetic data also carries advantages in usability: For instance, our toolset allows to validate robotics applications without *access* to the respective workcell. This includes off-site validation or the a-priori validation of applications at the planning stage of a robotics facility. Neither of these scenarios is possible with real-world data.

Its algorithmic nature gives our toolset further advantages. Most notably, the *time* needed to perform validation or to generate a validation sequence scales with available hardware. This is not possible with real-world validation data, which demands fixed time to record a single movement sequence. To verify our claim, we measured an example speed-up using our toolset on a fixed hardware configuration, an Intel Core i7 with 16 GB RAM and an NVIDIA GeForce GTX 1060 graphics board. Given a set of objects, workers and behaviours, we are able to estimate how many synthetic data sets our framework can generate in the same time a single real-world run would need. Table 2 sums our results. Notably, at a rate of 10 Hz, we can generate six synthetic multi-camera sequences in the same time a single real-world sequence takes to record. Finally, please note that our experiment does not consider the (negligibly small) time required to start the toolset and to place a few animation nodes for the human worker.

Closely related to time benefits are *coverage* benefits: Due to the synthetic and physically-based nature of our animations, we can easily generate additional validation sequences. Examples include permutation testing of behaviours over a fixed path or fuzzy testing with randomized animation nodes, segments, and

Table 2: Time benefits of synthetic validation data

frequency [Hz]	5	10	30	60
speed-up to real-time	13.2	6.6	2.2	1.1

behaviours. With real-world data, each of these automated runs would require a separate recording pass in the workcell.

Repeatability and *Modification* are two closely related points of concern: In the real world, exactly repeating a recorded motion sequence is mostly impossible and it is likewise difficult to partially adapt a real-world motion sequence after recording. However, both of these features are useful for validation, e.g. for testing occlusions or different camera positions. Our toolset, in contrast, can readily repeat motion sequences and partial motion segments can intuitively be adjusted. Finally, validation on synthetic data avoids certain risks of real-world validation. *Hazardous* cases, in particular, must be validated rigorously (e.g. to avoid human-robot collisions), which is not an option for real-world data.

Animating virtual characters with a physics simulation has several benefits when compared to an animation based on motion captures. One major drawback of motion captured animation is that motion retargeting is difficult even for slight adjustments to the original recording. Consequently, pre-defined motion trajectories either have to be fine-tuned by hand in a time-consuming process, or motion trajectories must be re-recorded for the desired motion. Algorithms for physics-based human animation instead are able to handle walking on uneven terrain [24] and can adjust their motions to grasped objects [4]. In the end, we decided to utilize the approach of [4] because of possible adaption to changes in the topology of the animated character, in obstacles and in grasped objects.

4 Experimental Evaluation

In this section, we validate our toolset for validation: We generate an animation sequence of a human worker in a workcell with our toolset and pass this sequence into a prototype monitoring system for shared human-robot workspaces.

4.1 Monitoring and Path Planning

The choice of monitoring and path planning systems distinctly enforces the form of validation data. Thus, we provide a short overview over both systems we wish to validate with our toolset. Our monitoring system (see [22]) accepts incoming images from an intrinsically and extrinsically calibrated multi-camera network. From these images, we derive a foreground-background segmentation (optionally, as non-binary confidences). We then merge resulting silhouettes over all cameras through an efficient and precise variant of the visual hull to derive a conservative approximation of workspace occupation. The subsequent path planning system (see [23]) incorporates multiple collaborating blackboard agents. These agents evaluate the approximated workspace occupation to suggest either collison-free or risk-minimized paths for the robot manipulator.

4.2 Validation Data

Our example animation consists of a human worker who starts at the front-right corner of the robot cell and walks to the back-right corner while waving her hand. We effortlessly designed the animation with our toolset and rendered the animation both as photo-realistic and silhouette images through a virtual multi-camera system of eight ceiling-mounted Full HD cameras. We then fed resulting images into the monitoring system and our path planning system. Finally, we compared monitoring results with exact ground-truth to evaluate monitoring.

4.3 Results

Over the course of the validation sequence, the monitoring system builds an approximation of the workspace occupation. We start by investigating effects of occlusions and conservative reconstruction compared to implicit ground-truth for human workers and other obstacles. To this end, we pass synthetic ground-truth silhouettes into the monitoring system. This procedure causes the approximated occupation to exceed the ground-truth occupation by an average of about 32% in volume. Depending on exact positions of human and robot, this occasionally makes the path planner discard occluded but otherwise viable path suggestions. Average movement times of the robot increase by an estimated 13% because occlusions occur mainly outside the usual robot paths. As a consecutive step, we feed photo-realistic virtual sensor images into the monitoring system. The monitoring system then performs foreground-background segmentation on the incoming images. Respective errors cause false-negative and false-positive classi-fication of workspace volumes. While workspace volumes grow by only 8%, robot paths slow down considerably with a 19% decrease in robot speed. This relates to the fact that noise artifacts, unlike occlusions, appear at dislocated positions in the robot workspace and have a more prominent impact on path planning. One can thus conclude that suppressing noise (e.g. by knowledge refinement) is more important than reducing occlusions (e.g. through camera optimization). Finally, note that both evaluations and the subsequent conclusion have only been possible due to availability of implicit ground-truth for synthetic test data.

5 Conclusion

In the preceding, we have presented our contribution towards validating robotics applications. Our toolset facilitates generating human animations: Users define animations intuitively by placing animation nodes in a workcell. Our toolset then automatically finds connecting animation segments and allows users to specify per-node or per-segment behaviour for human agents. Together with implicit ground-truth, the resulting, photo-realistically rendered and physically-based animations act as input to virtual validation of monitoring systems and path planners. We have discussed the advantages of our process, including a speedup over real-world validation, implicit and exact ground-truth, and incremental ad-justments to existing motion sequences.

In future work, we intend to increase the number of characters available per simulation to model collaboration between multiple humans and robots. Other goals include support for a wider variety of simulated sensors (e.g. for scenarios as in [19]), virtual online validation, and more rigid automated validation.

Acknowledgements
This work has partly been supported by the Deutsche Forschungsgemeinschaft (DFG) under grant agreement He2696/11 SIMERO.

References
1. Actin Software. www.energid.com/software/actin-simulation
2. Bullet Physics Library. www.bulletphysics.org
3. S. Carpin. USARSim: a robot simulator for research and education. ICRA, 2007.
4. S. Coros, P. Beaudoin, M. Van de Panne. Generalized biped walking control. ACM Transactions on Graphics 29.4, 2010.
5. R. Diankov, J. Kuffner. Openrave: A planning architecture for autonomous robotics. Robotics Institute, Pittsburgh, PA, Tech. Rep. CMU-RI-TR-08-34 79, 2008.
6. W. Donnelly, A. Lauritzen. Variance shadow maps. Interactive 3D graphics and games. ACM, 2006.
7. M. Freese et al. Virtual robot experimentation platform v-rep: A versatile 3d robot simulator. Simulation, modeling, and programming for autonomous robots, 2010.
8. S. Hargreaves, M. Harris. Deferred shading. Game Developers. Vol. 2. 2004.
9. P.E. Hart, N.J. Nilsson, B. Raphael. A Formal Basis for the Heuristic Determination of Minimum Cost Paths. Systems Science and Cybernetics SSC4, pp. 100-107, 1968.
10. N. Koenig, A. Howard. Design and use paradigms for gazebo, an open-source multi-robot simulator. IROS, 2004.
11. S. Lemaignan et al. Human-robot interaction in the MORSE simulator. HRI, 2012.
12. T. Möller, E. Haines, N. Hoffman. Real-Time Rendering, Third Edition. ISBN 9781568814247, AK Peters/CRC Press, 2008.
13. Newton Dynamics. www.newtondynamics.com
14. Open Dynamics Engine (ODE). www.ode.org
15. Open Source 3D Graphics Engine (OGRE). www.ogre3d.org
16. M. Olivier. Cyberbotics Ltd. Webots: professional mobile robot simulation. Advanced Robotic Systems 1.1, 2004.
17. M. Piccardi. Background subtraction techniques: a review. Systems, Man and Cybernetics, Vol. 4, 2004.
18. RoboDK. Sim. for industrial robots and offline programming. www.robodk.com
19. J. Thieling, J. Rossmann. Virtual Testbeds for the Development of Sensor-Enabled Applications. Montage Handhabung Industrieroboter. Springer Vieweg, Berlin, Heidelberg, pp. 23-32, 2017.
20. Vortex Studio. www.cm-labs.com/vortex-studio
21. T. Werner. Integration of an Interactive Raytracing-Based Visualization Component into an Existing Simulation Kernel. Master Thesis. University Bayreuth. 2011.
22. T. Werner, D. Henrich. Efficient and precise multi-camera reconstruction. Distributed Smart Cameras. ACM, 2014.
23. T. Werner, D. Henrich, D. Riedelbauch. Design and Evaluation of a Multi-Agent Software Architecture for Risk-Minimized Path Planning in Human-Robot Workcells. Montage Handhabung Industrieroboter. Best Paper Award.
24. J. Wu, Z. Popovic. Terrain-adaptive bipedal locomotion control. ACM Transactions on Graphics (TOG) 29.4, 2010.

Concept of Distributed Interpolation for Skill-Based Manufacturing with Real-Time Communication

Caren Dripke[1], Ben Schneider[2], Mihai Dragan[1], Alois Zoitl[2], Alexander Verl[1]

[1] Institute for Control Engineering of Machine Tools and Manufacturing Units (ISW)
Seidenstr. 36, Stuttgart, 70174, Germany, Tel. +49-711-685 84500
Caren.Dripke@isw.uni-stuttgart.de

[2] fortiss GmbH, An-Institut Technische Universität München
Guerickestr. 25, München, 80805, Germany, Tel. +49 89 3603522 0
schneider@fortiss.org

Abstract. Distributed manufacturing unit control can be implemented by equipping all manufacturing components with individual controls offering standardized skills. The control of multi-component groups often requires real-time communication and a communication architecture that is adapted to the distributed control concept. We present applicable communication concepts for distributed interpolation, where the distributed interpolation use case in particular demonstrates challenges caused by the synchronous execution of skills.

Keywords: distributed interpolation, deterministic real-time communication, time-Sensitive Networking, IEC 61499

1   Introduction

Fully networked and in real-time communicating automation systems, consisting of intelligent automation components - this vision is becoming more and more realistic in the age of digitization and industry 4.0. Controls, actuators and sensors can now be integrated into components because of the latest progress in electronics miniaturization. Embedded software makes of such components intelligent, communicative modules. These offer their functionality in the form of manufacturer-independent standardized automation functions called skills. Embedded software enables such components to take responsibility of their internal processes. Each component encapsulates the inner workings - all sensors and actors are controlled by integrated control - and offers abstract skills via a communication network. This signifies the integration of physical automation devices into the software control system [13] and is a step further towards the scientific goal of designing and building intelligent machines [18]. This further eliminates the need for classical wiring to a control cabinet and adjustment of signals/currents to specific protocols in order to interact with the mechanical component. The result is a component-based automation with a component-oriented engineering.

© Springer-Verlag GmbH Deutschland, ein Teil von Springer Nature 2018
T. Schüppstuhl et al. (Hrsg.), *Tagungsband des 3. Kongresses Montage Handhabung Industrieroboter*, https://doi.org/10.1007/978-3-662-56714-2_24

One research question is how component skills and their modularization can be (synchronously) executed in real time. A work package of the DEVEKOS research project aims to offer a communication solution for the envisioned component-based architecture. Goals are the continuous network-wide call for skills and the simple integration of third-party or previously unknown components with a focus on fault management, authentication, data confidentiality, and security.

DEVEKOS is funded by the German BMWi (Federal Ministry for Economic Affairs and Energy) with the goal to establish a consistent engineering as well as a workflow for safe, distributed and communicating multi-component systems in manufacturing units in the Industry 4.0 / Internet of Things (IoT) context [2]. The project focuses on the manufacturing unit and its components and does not specify solutions for distributed, decentralized manufacturing control or project planning. Integration into overlying levels of manufacturing execution systems (MES) and enterprise resource planning (ERP) systems are considered by using industry standards for information exchange as Automation-ML and OPC UA. DEVEKOS is looking to create concepts for engineering, safety, real-time communication and motion functionality that are easily applicable in industry.

The identified challenges of a communication concept have been presented in [6]. It is necessary to define a communication architecture that allows for real-time communication. Definition of the exchanged messages should aim for a compact representation of the information, including time stamping. Other aspects discussed were possible subnetwork structures as well as a handling strategy for possibly interrupted or corrupted communication. In this paper we introduce distributed interpolation as a use case to verify a concept that allows for real-time communication between individual components in the component-based architecture. We concentrate on real-time and deterministic communication as well as on the definition of compact messages. The definition of subnetwork structures is not in the scope of this paper.

2 State of the Art

We propose a concept of individually controlled components that interact with each other in order to fulfill an externally defined goal. The definition of Holonic Manufacturing Control (HMC) [13] can be applied to the skill architecture. HMC defines single components as interacting holons as well as groups that are made up of multiple components. This modularization has to be reflected on the software and communication architecture level. HMC allows to (re)combine, extend, amend or replace components or modules freely, which also implies publicly available interfaces that rely on industry standards [9]. Each holon offers skills to its environment, the offered skills depend on the level of the holon. A skill could be *OpenGripper()* for a holon describing a low-level gripper or *ManufactureWorkpiece()* for the uppermost holon level describing the manufacturing unit itself [10]. The high-level skill is itself a composed skill, meaning it is constructed by a sequence of lower-level skills. This orchestration of these lower-level skills and especially their synchronized execution is the main focus of this paper. This concept advances the manufacturing unit towards a distributed, skill-oriented architecture, where submodules can be exchanged arbitrarily without loss of the general function of the manufacturing unit or the need for complex changeover processes. Components from various manufacturers are expected to work together in groups i.e. to connect and communicate with each other. The need to adhere to different protocols or setups depending on the manufacturer is replaced by universal skills. Up until now, an HMC implemented as a multi agent system has only been used in production process planning [3], where communication cycle times are significantly longer. When even the lowest-level components like axes, grippers or sensors are integrated into an HMC, real-time communication means cycle times below 10 milliseconds [7] for their assigned skills.

Distributed control concepts have been introduced for programmable logic controllers (PLCs) within the IEC 61499 [5]. 4diac [8] is an IEC 61499 compliant development environment for modeling distributed industrial automation systems. It simplifies the engineering workflow and supports different communication patterns and protocols including real-time and deterministic communication by a recently developed Time-Sensitive Networking (TSN) [19] communication layer described in Section 3.

TSN is currently standardized by the Time-Sensitive Networking Task Group which is part of the IEEE 802.1 Working Group. The advantages of TSN can be summarized as follows:

- **Real-time:** TSN guarantees real-time capabilities by bandwidth reservation and configured schedules
- **Determinism:** TSN guarantees the transmission of packets by bandwidth reservation
- **Interoperability:** standard Ethernet as key enabler for Industrial Internet of Things (IIoT), as the need for dedicated fieldbuses is decreasing
- **Convergent network:** real-time critical and uncritical traffic are transmitted on the same cable, which results in better bandwidth utilization and reduction of cabling effort

The TSN substandard 802.1Qbv defines the allocation of Virtual Local Area Network (VLAN) to tagged and encoded priority values [16]. These allow simultaneous support of scheduled traffic, credit-based shaper traffic and other bridged traffic over Local Area Networks (LANs), thus making TSN a key enabler for the advantages mentioned above.

Fieldbuses and industrial Ethernet solutions such as EtherCAT, PROFINET or Sercos III are evaluated for their compatibility with the project requirements. EtherCAT is a real-time-capable Ethernet-based fieldbus developed at Beckhoff Automation. The EtherCAT approach makes the connected devices only read data addressed to them, whilst each communication telegram continues through the device. Similarly, input data are inserted while the telegram passes through. The telegrams are only delayed by a few nanoseconds, making this solution one of the fastest available. The bus system offers access speeds similar to local I/Os. EtherCAT enables not only the position control loop, but also the velocity control loop or even the current control loop to be closed via the bus, resulting in cost-effective drive controllers [1]. PROFINET IO is based on Switched Ethernet full-duplex operation and a bandwidth of 100 Mbps. It therefore provides the possibility of real-time communication with reserved bandwidth in its IRT-mode (Isochronous Real Time). In this case, a "sync master" transmits a synchroniza-tion message to which all "sync slaves" synchronize themselves with a synchro-nization accuracy of less than one microsecond [17]. SERCOS is a digital motion control bus that interconnects drives, motion controls, I/O, sensors and actua-tors for numerically controlled machines and systems. Launched in the 1990s, it is utilized for automation applications with high requirements for dynamics and precision. SERCOS III is the open, IEC-compliant third-generation SERCOS interface enabling persistent, bi-directional motion control communications in real-time between all drives and the motion control [11].

The innovative communication approach of TSN is also considered here as a potential solution, as it could be suitable for a consistent and reliable com-munication in distributed, asynchronous, event-driven and service-oriented ar-chitectures. Future work within the communication work package of this project includes the evaluation of known solutions, followed by the development of a suitable communication solution and an experimental validation of this concept.

3 Applicable Communication Proposal

The afore mentioned challenges for distributed interpolation imply several com-munication requirements [6], mainly real-time and deterministic traffic on a con-vergent network as described in Section 1. There are plenty of already existing real-time-capable, proprietary fieldbuses on the market, as described in Section 2. These industrial protocols are not enabling an Industrial Internet of Things (IIoT). Reasons are the limited scaling of bandwidth to 100 Mbps, the limited interoperability due to special hardware and the lack of supporting mixed traffic on a single network cable. TSN meets all these requirements the other fieldbuses lack, as described in Section 2, and is therefore chosen as the communication protocol for this project.

The proposed communication pattern is 'publish/subscribe' (pub/sub). It is best suited for an asynchronous, event-driven and distributed architecture with real-time requirements. Additionally, it enables multicast communication. The 4diac framework already provides an UDP based communication layer for pub/sub data streams with TSN support, which was implemented in the BEinCPPS project. This communication layer adds virtual local area network (VLAN) and priority configuration data to IEC 61499 compliant publish Service Interface Function Blocks (SIFB). The ID data input parameter of publisher SIFBs is used to select a specific network stack and to configure the parameters.

The ID parameter of 4diac's TSN communication layer has the following format: `fbdk[].tsn[<ip>:<port>:<vlan_id>:<prio>]`. Whereas `fbdk[]` is re-sponsible for the appropriate ASN.1 encoding defined in IEC 61499-1 Annex F [5], `tsn[..]` specifies that the TSN enabled communication layer of the 4diac runtime environment is used.

There are different message types necessary for the current implementation of distributed interpolation. The proposed TSN approach maps the message types to priorities (0 highest, 7 lowest priority) and VLANs which are then treated differently according to the configuration of the TSN schedule. The different message types and priorities are mapped as follows:

- MSG 1: highest priority (0) - represents a skill request;
- MSG 2: priority (1) - for synchronization, negotiation and interpolation tasks;
- MSG 3: no priority - device internal data for motion control;
- Network control traffic: priority (2) - IEEE 1588 Precision Time Protocol (PTP) [15] for time synchronization;
- Best effort network traffic: lowest priority (7).

MSG Type	Sender Addr.	Target	State Status	Max. Reach. Pos.	Current Pos.
MSG 1	4	max. 64	1	—	—
MSG 2	4	max. 64	1	1	4
MSG 3	4	max. 64	1	—	—

Table 1: Different message types and their tags including the size [Byte]

MSG 1 through 3 are directly connected to the distributed interpolation and will be presented in detail in the following chapter. The separation into three main message types is motivated by the fact that the skill request (MSG 1) should be a network-wide standard. The standardization process what a skill request message entails is not yet concluded. Changes are to be expected but will not affect the underlying messaging concepts. The separation between MSG 2 and MSG 3 is motivated by the distinction between internal messaging (on the individual axis control) and the negotiation between multiple axis controllers via the network. The preliminary message sizes are shown in Table 1. An address and a transmitted target (requested skill, possibly subdivided in tasks) as well as a communication status are always included. In case of a negotiation in the interpolation task, an axis will share its maximum reachable position as an offer, as well as its current position.

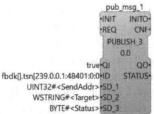

Fig. 1: TSN enabled publish SIFB in 4diac

The appropriate TSN enabled network SIFB can be derived when message type, priority mapping, VLAN mapping and size of the message are defined, see Figure 1. The example ID parameter fbdk[].tsn[239.0.0.1:48401:2:0] of a publisher SIFB would map a message stream of the proposed interpolation MSG 1 to the UDP multicast group with IP address 239.0.0.1 on port 48401 to VLAN 2 with priority 0. The data types are chosen according to the different tags of the message (e.g., sender address, target, ...) and their size. The implementation of the TSN layer and the analysis of message types and their mapping to TSN priorities and VLANs will be the base for evaluating the communication concept.

4 Architecture for Distributed Interpolation

Distributed skill-based control brings communication challenges that are not covered by current centralized control communication architectures. Functionalities like synchronization, online adaptations to the motion target or status notifications are currently requests from the manufacturing network. These functionalities will be encapsulated in skills in the component based architecture. The first advances towards distributed control as IEC 16499 [5] are mostly used to implement sequential event-chains of events. Non-sequential skill request however, as e.g. the movement of multiple axes interpolating a specified path, are not specified. Here, the independently controlled individual movements (individual task decomposition of the skills has yet to be defined) have to be synchronized. If any discrepancy from the planned path is detected, all axes have to determine a compensation strategy (adapt motion target, abort movement, ...). Communication offers the possibility to solve such tasks in collaboration [18]. It is expected, that a communication architecture that meets the requirements of the distributed interpolation use case is transferable to any upper-level HMC / multi-component group solving non-sequential skill requests. The orchestration [12] of skill-execution or computation of transformations, kinematic relations or dynamic capabilities are not discussed here but will be subject of a future publication.

The distributed control architecture proposed requires all components to be ablt to interpolate motion have identical communication interfaces. The implementation is realized with three state machines (SM) which separate the modular responsibilities as follows (see also Figure 2):

- Interpolation Master (IPO-Master): This SM is available in all component controls, but for each multi-component group only a single interpolation master is chosen. In this context, we assume the choice of the IPO-Master as given (could e.g. be done by a bidding algorithm, by fastest response times,...). The IPO-Master receives the skill request in Message Format 1 (MSG 1). It passes all relevant information on to the Communication State Machines (CommSM) of all axes in the group via MSG 2. The coordination and execution of all motion, synchronization and negotiation until an agreement is responsibility of the IPO-Master.
- Communication State Machine: Checks incoming requests and negotiates with IPO-Master if necessary. Here, high loads of messages are possible and time-crucial. Passes on the agreed motion to MotionSM via MSG 3.

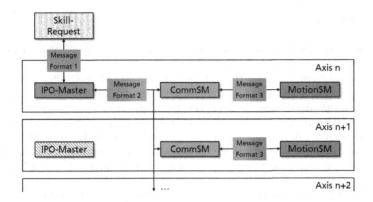

Fig. 2: Message-based communication between the state machines (SM): An IPO-Master is chosen, receives skill request (MSG 1). IPO Master communicates with the CommSMs through MSG 2. Communication between MotionSM and CommSM is internal over MSG 3.

– Motion State Machine (MotionSM): Executes motion, not accessible directly, but only communicates with the CommunicationSM via MSG 3.

An axis under a distributed interpolation HMC offers skills such as *GoToPosition()* and other motion profiles. The proposed set of skills are still under discussion, but are oriented to the PLCOpen [14] and CANOpen [4] standards. A combination of these motion skills then offers the possibility to combine them in a multi-component group to orchestrate interpolated motion.

5 Outlook and Conclusion

We presented the context of the DEVEKOS research project. The proposed communication of distributed interpolation can present an example for other non-sequential multi-component skill-requests. Challenges of the current communication concepts and possible technologies were also presented. The next step is the realization of the concept with a prototype using 4diac and TSN.

6 Acknowledgements

The DEVEKOS research and development project is funded by the Federal Ministry of Economic Affairs and Energy (BMWi). It is part of the technology program "PAICE Digitale Technologien für die Wirtschaft" and is guided be the DLR Project Management Agency "Information Technologies/Electromobility", Cologne. The authors are responsible for the contents of this publication.

The present work has been partially carried out in the framework of the BEinCPPS project, which has received funding from the European Union's Horizon 2020 research and innovation programme under grant agreement No 680633.

References

1. Beckhoff: Beckhoff information system. https://infosys.beckhoff.com/index_en. htm (2017) accessed 27.10.2017.
2. Bundesministerium für Wirtschaft und Energie: DEVEKOS. http://www. digitale-technologien.de/DT/Redaktion/DE/Standardartikel/PAICEProjekte/ paice-projekt_devekos.html (2017) accessed 06.04.2017.
3. Bussmann, S., Jennings, N.R., Wooldridge, M.J.: Multiagent systems for manufacturing control: a design methodology. Springer Science & Business Media (2004)
4. CAN in Automation (CiA) e. V.: Drives and motion control device profile; Part 2: Operation modes and application data. Draft Standard Proposal No. 402 (10 October 2016)
5. DIN EN 61499-1: Funktionsbausteine für industrielle Leitsysteme – Teil 1: Architektur (2013)
6. Dripke, C., Verl, A.: Challenges in distributed interpolation with multi-components systems in future-oriented manufacturing units. Proceedings of the 47th International Conference on Computers and Industrial Engineering 2017 (2017 - in press)
7. Dürkop, L.: Automatische Konfiguration von Echtzeit-Ethernet. Springer (2016)
8. Eclipse: 4diac Framework for Distributed Industrial Automation and Control. https://www.eclipse.org/4diac/index.php (2007) accessed 19.10.2017.
9. Frey, K.: Offene Steuerungssysteme-eine Zwischenbilanz. In: FTK'97. Springer (1997) 271–297
10. Helbig, T., Henning, S., Hoos, J.: Efficient engineering in special purpose machinery through automated control code synthesis based on a functional categorisation. http://link.springer.com/content/pdf/10.1007%2F978-3-662-48838-6_9.pdf (2016)
11. Hibbard, S.: Industrial ethernet: The key advantages of sercos iii. https://www. automation.com/pdf_articles/Rexroth_SERCOS_III__L-1.pdf (2010) accessed 27.10.2017.
12. Jammes, F., Smit, H., Lastra, J.L.M., Delamer, I.M., eds.: Orchestration of service-oriented manufacturing processes. Volume 1., IEEE (2005)
13. Leitão, P.: Agent-based distributed manufacturing control: A state-of-the-art survey. Engineering Applications of Artificial Intelligence **22**(7) (2009) 979–991
14. PLCopen: TC2 - Motion Control. http://www.plcopen.org/pages/tc2_motion_control/ (2011) accessed 30.10.2017.
15. Precise Networked Clock Synchronization Working Group: Enhancements for scheduled traffic (2008)
16. Precise Networked Clock Synchronization Working Group: Ieee standard for a precision clock synchronization protocol for networked measurement and control systems (2008)
17. Profinet: Profinet system description (2012)
18. Shen, W., Norrie, D.H., Barthès, J.P.: Multi-agent systems for concurrent intelligent design and manufacturing. CRC press (2003)
19. Time-Sensitive Networking Task Group: Ieee standard for local and metropolitan area networks – bridges and bridged networks – amendment 25: Enhancements for scheduled traffic (2015)

Service-oriented Communication and Control System Architecture for Dynamically Interconnected Assembly Systems

Sven Jung[1,a], Dennis Grunert[1,b] and Robert H. Schmitt[1,2,c]

[1] Fraunhofer Institute for Production Technology IPT,
Steinbachstr. 17, 52074 Aachen, Germany
[2] Laboratory for Machine Tools and Production Engineering (WZL) of
RWTH Aachen University, Campus-Boulevard 30, 52074 Aachen, Germany
[a]sven.jung@ipt.fraunhofer.de
[b]dennis.grunert@ipt.fraunhofer.de
[c]r.schmitt@wzl.rwth-aachen.de

Abstract.
Varying sales expectations, shorter product life cycles, and a rising product variability caused by individualization: Manufacturing companies are facing the challenge of continually adapting their assembly systems to these and other dynamic conditions. The dynamic interconnection of stations enables to design flexible and individual assembly sequences for each product. Benefit is the possibility to assemble customer-specific products and to react to dynamic conditions of the system. This paradigm shift entails a raise in the level of complexity and flexibility in terms of coordinating the different product flows, exceeding the possibility of existing control systems. Within this paper, we therefore present the concept of a service-oriented communication and control system architecture as a solution to cope with the introduced challenges and requirements of Dynamically Interconnected Assembly Systems.

Keywords: Manufacturing System, Assembly, Agile control.

1 Introduction

Conventional organizational forms of assembly systems, such as line assembly where products are assembled in a fixed process sequence on rigidly linked stations, come to their limits due to their inflexible design. Reconfigurations cause high efforts regarding time and costs. Thus, there is an emerging demand for more flexible and scalable approaches that allow for adaptable and reconfigurable systems [1].

A promising solution are Dynamically Interconnected Assembly Systems (DIAS), defined as "An assembly system is dynamically interconnected if it provides a flexible assembly sequence for each individual product (job route), without any limitations in time and space" [2]. By decoupling stations from each other this so-called job route not

© Springer-Verlag GmbH Deutschland, ein Teil von Springer Nature 2018
T. Schüppstuhl et al. (Hrsg.), *Tagungsband des 3. Kongresses Montage Handhabung Industrieroboter*, https://doi.org/10.1007/978-3-662-56714-2_25

only allows to assemble customer-specific products up to lot size 1, but also to individually react with quick adjustments to dynamic conditions of the system, like system reconfigurations, failures, or assembly resource availabilities. Figure 1 depicts different types of the resulting job routes. Different products can be assembled on the same facility using individual process sequences (blue and orange). Depending on the status of the system, job routes can be rescheduled (blue and light blue). By adding or removing assembly resources (red), scalability in terms of avoiding bottlenecks and introducing new products is achieved.

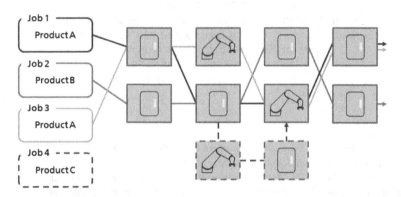

Fig. 1. Illustration of different types of flexible and individual process sequences (job routes) in Dynamically Interconnected Assembly Systems (see [2]).

The control system is responsible for planning the products to be assembled, distributing them to all available assembly resources, and controlling and monitoring the execution in real-time. With the introduction of Dynamically Interconnected Assembly Systems, new requirements emerge. Central task in a paradigm shift like this is to efficiently coordinate the different product flows and monitor the status and abilities of the assembly resources. The level of complexity and flexibility substantially exceeds the possibilities of existing control systems. Therefore, it is necessary to develop new approaches for the architecture of such systems.

In this paper, we aim to define a reference architecture for control systems of individualized and flexible production environments. Functioning as an idealized model with predefined structures and types as well as their interactions and responsibilities, specific systems can use these structures as a basis and add further system-dependent details. First, the state of the art of control system architectures is presented and requirements regarding more flexibility and scalability are identified. Afterwards, the concept of a service-oriented communication and control system architecture for Dynamically Interconnected Assembly Systems meeting these challenges and requirements is introduced. Finally, a conclusion and an outlook is given.

2   State of the Art

Manufacturing Execution Systems (MES) can be seen as an intermediate layer to coordinate between planning enterprise systems and executing systems of the machine controls [3] (see Fig. 2). One central component is the detail planning and control of the shop floor. However, the complexity and flexibility of Dynamically Interconnected Assembly Systems exceed the possibilities of typical process control systems regarding individual and variable process sequences for products and the consideration of abilities and states of resources for routing [2]. Even though PRS Technologie GmbH offers the software system synchroTecS, allowing forming and schedule batches for a clocked individual production [4], it lacks in adapting process sequences of individual products in real time. The Fraunhofer Institute for Production Technology IPT developed a quickly reconfigurable control system with individually composable process sequences for the automated production of stem cells [5]. A dynamic scheduling of the required process sequence to suitable manufacturing resources according to the status and abilities of the system is not yet implemented. KIMEMIA and GERSHWIN proposed an algorithm for the dynamic control of Flexible Manufacturing Systems (FMS) considering the system status [10]. Even if this approach focuses the scheduling of production rates instead of individual process sequences and neglects re-configurability, there are already dedicated computation modules for off-line planning, production rates determination, online routing calculation, and execution control.

Aim is to move from an individual software development towards intelligent systems adapting independently to the needs of the facility [6]. Cyber-Physical Systems (CPS) suggest the structure of loosely coupled networks of cooperating autonomous agents to achieve a reconfigurable and flexible manufacturing control [7] (see Fig. 2).

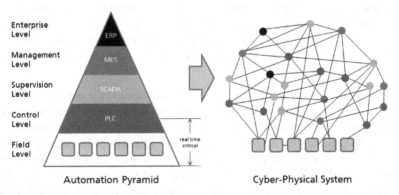

Fig. 2. Supersession of the hierarchical automation by loosely coupled networks with distributed services (adapted from [7]).

For this, Plug and Produce forms the basis to integrate new stations or abilities to the system [8]. MENDES et al. suggest the use of a Service-oriented Architecture (SoA), which provides control logic and functionalities of components encapsulated in software modules that respond to specific requests [11]. This approach allows to dynamically compose complex process sequences using temporally and spatially distributed abilities. Main challenge is to uniformly implement a system-wide information model, since every domain and vendor not only has its own protocols but especially individual ways of describing the abilities and information. A promising approach are modular abstraction layers between execution system and hardware, which encapsulate the diversity of interfaces and provide a uniform way of communicating internally in the control system [9].

3 Architecture Design

There is yet no overall communication and control system architecture considering all these different aspects to solve the challenges of Dynamically Interconnected Assembly Systems. Especially, the need for an easy re-configurability of the control system and a temporal and spatial separation of the process sequence and final route resulting from the demand for individual and dynamically adapting routes is not yet addressed.

Most control systems are large and complex fulfilling the purpose of planning, managing, and controlling the production. Interactions and responsibilities of entities have to be chosen in a way to maximize flexibility whilst minimizing the overhead for interdependencies. Therefore, the communication and control system architecture for Dynamically Interconnected Assembly Systems is divided into a number of different subsystems cooperating via open and global information networks (see Fig. 3). Each subsystem has its own distinct role and itself consists of various modules.

Planning System. The planning system covers the business logic and long-term planning of the whole enterprise. It consists of two main modules relevant for the control system. The Order Management provides order information, like the products to be assembled and the different variants. The Product Management manages product features, like required dimensions or quality levels, and the respective process sequences.

Execution System. Central subsystem of the architecture is the execution system. Its primary task is to calculate an optimal job route for the assembly of each product and to manage the execution. For this, all conditions and restrictions of the assembly system are taken into account, like the status and abilities of the available assembly resources. As already discussed, it is necessary to decouple the single process step from the actual temporal and spatial execution. Therefore, the role of the execution system is divided into three consecutive modules: Matching, Routing and Assembly Control. In a first step, the Matching calculates all possible job routes for a particular product. This is achieved by matching the required sequence of assembly steps to all available resource abilities in a direct comparison of the product and process features. Exemplarily, the adjustment of the maximum available space of the executing station with the maximum size of the product can be mentioned. In case of a reconfiguration of the assembly system, the Matching independently ensures to update the sets of possible job routes. Next, in case of an order, the Routing selects the most suitable route for the product from the respective set of all possible routes optimizing regarding some target figures. For this, it considers dynamic system conditions, like the status of the assembly resources or the priorities of orders. Target figures may vary depending on the company goals, use case, or state of the assembly system and could be a maximization of the resource utilization, minimization of the production costs, or adherence of deadlines, for example. This route selection is also independently updated to changing conditions. Finally, the Assembly Control is responsible to execute the currently selected job routes by activating assembly resources and the intralogistics system.

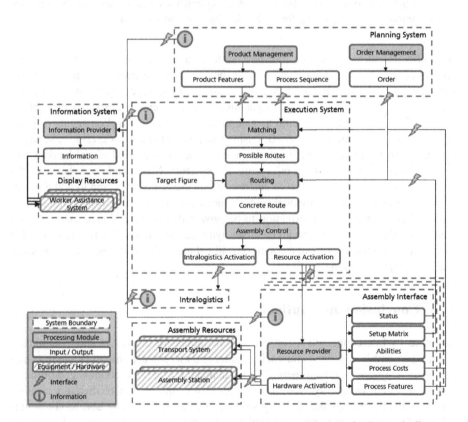

Fig. 3. Service-oriented communication and control system architecture for Dynamically Interconnected Assembly Systems

Information System. While the control systems operates at a technical level, the transparency of the overall system is indispensable for the human as a control and a possible intervening instance. Information systems take over the integration of the workers and thus form the interface between human and machine. For this purpose, the information system collects information from all the various systems. Central module is the Information Provider aggregating the information and distributing it to the individual worker assistance system depending on the worker and situation. Various end-user devices are conceivable, such as displays, smartphones, tablets, or smart wearables. This enables, for example, to provide individual assembly manuals for manual assembly tasks or to display the system status to operators.

Intralogistics System. Task of the intralogistics system is the organization and realization of the internal flow of materials in order to provide a punctual availability. This includes the material supply at the assembly stations, like general cargo, bulk material, and liquids, as well as the management and handling of goods. The transportation of products between single assembly stations is explicitly excluded, since this is regarded as an assembly resource in the form of a flexible transport system.

Assembly Resource. Flexible transport systems and assembly stations are the resources of the assembly system and provide the operational execution of the single process steps. Re-configurability of the control system is achieved by following the service-oriented approach enabling to dynamically add or remove resources. Therefore, assembly resources are considered as autonomous and self-controlled agents offering their functionalities and characteristics using independent and self-descriptive services. For this, a software middleware acts as mediator between the individual hardware or rather resource control and the other software systems. Main module is therefore the Resource Provider whose task is to abstract the interface diversity and to add a meaningful description to the individual resource according to the system-wide descriptive model. As discussed, the information to be provided by each resource may vary depending on the system. This could be either dynamic conditions, like the status needed by the Routing module, or static conditions, like the abilities with respective costs and features and the setup matrix.

4 Conclusion and Outlook

Dynamically Interconnected Assembly Systems allow for individual and dynamically adapting assembly sequences for each product and thereby are able to tackle the challenges of today's assembly systems, like the demand for highly individualized products and the resulting variability. Such a paradigm shift results in new challenging requirements for the control system. Existing solutions explicitly lack in an easy re-configurability and a temporal and spatial separation of the process sequence and final job route. Promising approach is to use a loosely coupled network of cooperating autonomous agents offering their data and abilities via service-oriented interfaces. According to this, a communication and control system architecture was developed and presented. Central approach is the service-oriented integration of assembly resources and the division of the role of the execution system into finding possible routes for a product, dynamically selecting the optimal route, and executing the process steps of the currently selected route. Result is a flexible and modular control system, which independently adapts to changing conditions without the intervention of experts. Next step is to implement the identified modules and algorithms, to define the required communication interfaces, and develop a descriptive model to describe and match abilities to product features.

Acknowledgement

This research and development project is funded by the German Federal Ministry of Education and Research (BMBF) within the "Innovations for Tomorrow's Production, Services, and Work" Program (funding number: 02P15A146) and implemented by the Project Management Agency Karlsruhe (PTKA). The author is responsible for the content of this publication.

SPONSORED BY THE

References

1. ElMaraghy, H., Schuh, G., ElMaraghy, W., Piller, F., Schönsleben, P., Tseng, M., Bernard, A.: Product variety management. CIRP Annals - Manufacturing Technology 62(2) (2013) pp. 629–652.
2. Hüttemann, G., Göppert, A., Lettmann, P., Schmitt, R.: Dynamically Interconnected Assembly Systems – Concept Definition, Requirements and Applicability Analysis. WGP-Jahreskongress 7(1) (2017) pp. 261-268
3. Verein Deutscher Ingenieure: VDI-5600 – Manufacturing Execution Systems (MES). Beuth Verlag, Berlin
4. Schmitt, R., Ellerich, M., Groggert, S.: Auf dem Weg zur Individualproduktion 4.0. VDI-Z Integrierte Produktion 157(5) (2015) p. 71
5. Kulik. M., Ochs, J., König, N., McBeth, C., Sauer-Budge, A., Sharon, A., Schmitt, R.: Parallelization in Automated Stem Cell Culture. CIRP Conference on BioManufacturing 65 (2017) pp. 242-247
6. Niggemann, O.: Industrie 4.0 ohne modelbasierte Softwareentwicklung. ATP edition (5) (2014) pp. 22–30
7. VDI/VDE-Gesellschaft Mess- und Automatisierungstechnik: Cyber-Physical Systems: Chancen und Nutzen aus Sicht der Automation. Verein Deutscher Ingenieure (2013), Düsseldorf
8. Reinhart, G., Krug, S., Hüttner, S., Mari, Z., Riedelbauch, F., Schlögel, M.: Automatic configuration (Plug & Produce) of Industrial Ethernet networks. IEEE/IAS International Conference on Industry Applications 9 (2010) pp. 1–6
9. Jung, S., Kulik, M., König, N., Schmitt, R.: Design of a Modular Framework for the Integration of Machines and Devices into Service-oriented Production Networks. WGP-Jahreskongress 7(1) (2017) pp. 167-174
10. Kimemia, J., Gershwin, S.: An Algorithm for the Computer Control of a Flexible Manufacturing System. IIE Transactions 15(4) (1983) pp. 353-362
11. Mendes, J., Leitao, P., Colombo, A., Restivo, F.: Service-Oriented Control Architecture for Reconfigurable Production Systems. IEEE International Conference on Industrial Informatics 6 (2008) pp. 744-749

Predictive Control for Robot-Assisted Assembly in Motion within Free Float Flawless Assembly

Christoph Nicksch, Christoph Storm, Felix Bertelsmeier and Robert Schmitt

Chair of Production Metrology and Quality Management, RWTH Aachen University,
52074 Aachen, Germany
c.nicksch@wzl.rwth-aachen.de

Abstract. This work focuses on the application of model predictive control (MPC) to the trajectory tracking problem for the integrated assembly of truck windshields in motion. Due to the continuous movement of the products, the handling device holding the windshield must be synchronized to the moving truck cabin to meet tolerance requirements. Using a MPC approach a model is derived to simulate the future system behavior to obtain a control law. The application of the control model is numerically simulated for effectiveness over short time periods for transient targets. The simulated results are experimentally verified on a full-scale demonstrator mimicking an actual assembly line environment. The experimental results show that the MPC approach is suitable for a windshield assembly in motion compensating system dead times and fulfilling synchronization between handling system and product. The presented approach allows for the efficient integration of automated assembly processes using state of the art handling systems into continuously moving assembly lines.

Keywords: Model Predictive Control, Large Scale Metrology, Assembly in Motion.

1 Introduction

Modern day automotive manufacturing is based on the principles of assembly line production, which is in constant motion to decrease throughput times [1]. Following this principle, the periphery such as handling systems need to be synchronized to the movement of the products. Especially for processes with narrow tolerances, the synchronization between handling and conveying system represents a major challenge. As a result, these processes cannot be integrated into the production flow. In particular in the field of automotive final assembly, this problem occurs when dealing with large and bulky workpieces. The complex shape of workpieces and, in addition, the narrow tolerances in the submillimeter range require process automation, which is currently only possible under stationary conditions [2]. Therefore, the products are stopped and are accelerated again to the velocity of the moving assembly line after the process. This causes additional buffer sections and, consequently, requires space on the shop floor [3]. As a result, productivity per unit time is decreased and additionally factory resources are strained [2].

© Springer-Verlag GmbH Deutschland, ein Teil von Springer Nature 2018
T. Schüppstuhl et al. (Hrsg.), *Tagungsband des 3. Kongresses Montage*
Handhabung Industrieroboter, https://doi.org/10.1007/978-3-662-56714-2_26

The synchronization between handling and conveying system comprises the fundamental motion control problem of trajectory tracking and can only be achieved by the use of feedback controllers [4, 5]. To achieve synchronization, dead times must be compensated predicting the future system behavior. Concerning this problem, the goal of this paper is to apply a model predictive control (MPC) approach to enable an assembly in motion for large components fulfilling synchronization between handling and conveying system. In detail, the approach must be capable of compensating system dead times and rejecting system disturbances such as vibrations of the conveyor and measurement noise.

First, synchronization principles for an assembly in motion are presented deriving the deficit and finally, the cause for this work. Then the design of the MPC follows. To validate the approach, first tests in a simulation environment and afterwards on demonstrator are carried out. The demonstrator shows a MPC for robot-assisted assembly in motion for truck windshield comprised of multiple input multiple output (MIMO) system using robot velocity for x, y and z as the input and the robot position as the output while the conveying system provides the target trajectory. The applied model describes the system dynamics such as the acceleration behavior. It is derived from a step response. The basic idea consists of using a metrology system as so-called global reference system obtaining sensor position data of both robot and conveying system [6]. By considering these inputs, the control algorithm must compensate the dead time of the metrology system to achieve synchronization between robot and conveying system. The final paragraph draws a conclusion and addresses open points for future works.

2 State of the Art

Three different ways to achieve synchronization for an assembly in motion are presently used [3, 5]: Mechanical synchronization, controlled synchronization and controlled synchronization with a feedback loop. The simplest way represents the mechanical synchronization. Components such as mechanical stops connect the handling device with the conveyor by force or form fit. Considering that neither vibrations through the mechanical connection, nor wear of the mechanical components are taken into account, the desired accuracy of the assembly process cannot be achieved. The principle of controlled synchronization determines the conveyor velocity by a sensor deriving the target velocity for the handling device. Also, this principle does not consider disturbances such as oscillations on the product or handling device. A position deviation between product and handling device cannot be detected so that this method is not suitable for tasks with narrow tolerances. The synchronization with a feedback controller allows to detect and compare the position deviation between handling device and conveyor. The conveyor position is usually provided by a position sensor while the robot position is determined by encoder values of the robot axis. Since disturbances on the product such as vibrations cannot be measured this approach is unable to meet narrow tolerances. The feedback controller approach must be enhanced to achieve higher accuracy. Therefore, disturbances must be measured right on the product and as close as possible to the end effector of the robot. In addition, the control algorithm must consider the dynamic behavior of the assembly process to reject disturbances and to compensate system dead times.

The MPC is a model-based control approach, which considers dynamic system behavior and is capable of compensating dead times. Current works show that there is a trend to use MPC for high dynamic systems as well as for robotic applications. RITSCHEL applies a MPC with a sampling rate of 4 milliseconds on an exhaust turbocharger. Considering the technical limits of the turbocharger by MPC constraints, the operating area can be enlarged to higher efficiency [7]. KÜHN ET AL. develop a MPC using successive linearization to navigate a mobile robot in real-time, which addresses the tracking problem in this paper and shows the effectiveness of MPC in the field of robotics [8]. The latest works on model predictive control emphasize that there is a high potential to realize an assembly in motion compensating dead times and disturbance using MPC to synchronize the movement between the handling device and the product.

Thus, it appears that the MPC is a suitable approach to solve the given problem in this work. The model predictive control uses a dynamic model of the process in the optimization problem to predict the future evolution of the process over a finite-time horizon to determine the optimal input trajectory with respect to a specified cost function [9]. MPC can account for process constraints motivated by economical or technical point of views. Taking this into account maximal velocities, accelerations and forces on workpieces, robots or conveyors can be set up to specify their technical limits. Furthermore, MPC can consider multivariable interactions enabling the control of MIMO systems.

3 Experimental Setup

The demonstrator consists of a six-axis robot (ABB IRB 4600) with a vacuum gripper to manipulate the windshield of a truck cabin placed on top of an automated guided vehicle (DONKEYmotion). Fig. 1 below shows the experimental setup with the gripper attached to the robot placing the windshield to the truck. During the experiment the automated guided vehicle (AGV) moves on a straight line providing the target trajectory for the robot. An indoor-GPS system (Nikon Metrology) is used following the principle of a global reference system to guide the robot towards the truck cabin [10]. The indoor-GPS (iGPS) as a large-scale metrology system is suited for positioning and tracking of large components with an accuracy down to 200 μm up to a distance of 80 meters [2, 10, 11, 12]. Several iGPS receivers can be merged together to build a frame gaining additional information about rotation angles. These frames are measured simultaneously with a frequency of 40 Hz building a data base for referencing different objects. The common tolerances are about 1.5 mm in the field of automotive final assembly.

To reference the gripper and the truck cabin, a frame of receivers is fixed on each ($Tool_{iGPS}$ and $Target_{iGPS}$). Considering that the frames are measured in the iGPS world coordinate system ($World_{iGPS}$) and the robot works in its own coordinate system ($Tool0_{Robot}$), the iGPS data must be transformed to the robot world coordinate system

(World$_{Robot}$). For this purpose, homogenous transformation matrices are used. The unknown matrices $_{Frame}H^{Tool0}$ and $_{iGPS}H^{Robot}$ must be determined by calibration using numerical optimization.

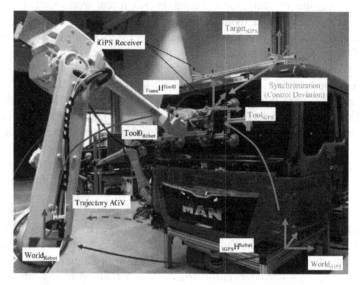

Fig. 1. Experimental setup for truck windshield assembly in motion

4   Model Predictive Controller

4.1   Design of the MPC

The goal of the devised model predictive controller is to synchronize the trajectory of the robot and follow the path of the truck cabin with the help of the coordinates provided by the iGPS. Fig. 2 shows the structure of the MPC. The truck cabin placed on the AGV provides the target values (w), while the robot as the controlled system provides the measured output (y) for the MPC. Depending on the current values of these variables, the MPC calculates the manipulated variable u using its internal model. The variable u sets the velocity of the robot in its tool coordinates for x- y- and z-direction. Due to the fact, that the iGPS causes dead times, both the dead time of the controlled system and the dead time of the reference trajectory must be compensated. The internal model of the MPC focuses only on the controlled system. Consequently, an additional dead time compensator (DTC) must consider the dead time of the reference trajectory separately. Taking this into account, a Kalman filter combined with a linear predictor modifies the reference trajectory for compensating dead time. Based on the equation of motion, the Kalman filter estimates both the position (p) and the velocity (v) of the truck cabin, which form the inputs for the predictor. It also increases the sample rate by linear interpolation to achieve a higher resolution and to evoke a more dynamic behavior of the controller. The linear predictor uses the estimated velocity to calculate a position correction in accordance to the equation of motion. The output of the predictor is position data in x, y and z and provides the input for the MPC (w). The given system considers a linear movement with constant velocity of the target object, which simulates a conveying system in real assembly line as a rough approximation.

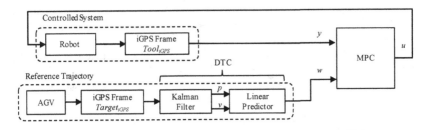

Fig. 2. Structure of the MPC

For the design of the MPC, the subsystems iGPS and robot have been investigated in detail. To set up the parameters of the Kalman filter, the dynamic standard deviation of the iGPS regarding the Cartesian position has been determined to 0.72 mm at a velocity of 100 mm/s, which represents the AGV's velocity during the assembly process. Also, the static standard deviation has been determined to 0.05 mm, which shows that iGPS loses accuracy under dynamic conditions (see [12]). To define the internal model of the MPC, the dead time is determined for iGPS (180 ms) and robot (50 ms). The dynamic behavior of the robot is modeled by black box testing. Velocities in x-, y- and z-direction on the robot tool are set as input values for the black box receiving iGPS position data of the robot tool as the output. In this way, a transfer function is determined to describe the dynamics of the robot. It was found out that there are no significant differences between transfer functions in x-, y- and z-direction moving the robot up to 100 mm/s, so that one transfer function is used for all Cartesian directions.

4.2 Results of the Simulation

The identified system parameters are applied to a simulation environment in MATLAB/Simulink according to the system structure shown in Fig. 2. A first performance analysis of the MPC is performed and absolute accuracies are obtained. A real-life assembly process is simulated involving target in motion and the robot tries to catch the target along a straight line with constant velocity. The sample rate of the MPC is set to 24 ms using a prediction range of 40 steps and a control range of 3 steps, which are found out to be a feasible configuration. Additionally, a velocity limit $u = 200$ mm/s is imposed as a boundary condition on the MPC for safety purposes.

The synchronization of the robot following a moving object is represented in Fig. 3 showing the control deviation as well as the trajectories. The robot starts from a stationary point, while the target point moves continuously with 100 mm/s. As shown in Fig. 3, synchronization in x-, y- and z-direction is achieved after 8 seconds. A slight overshoot of 2 mm appears while the robot reaches the setpoint in y-direction. A slight overshoot of 0.5 mm at 2 s can be seen. This behavior can be further reduced by a stronger penalization of the manipulated variables u (refer Fig. 2) taking into account the trade-off that the controller behaves slower.

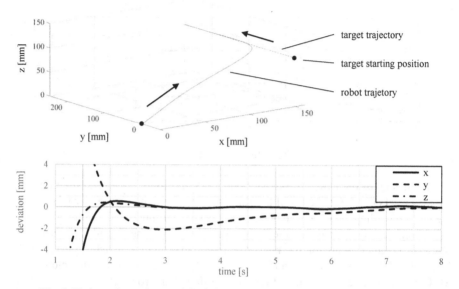

Fig. 3. Trajectories and control deviation between robot and moving target object

The simulation results show that the MPC is capable of synchronizing the robot to a target object and compensates the dead time of the subsystems. These insights obtained from the simulation results were used to experimentally validate the model.

5   Experimental Results

5.1   Reaching a Static Position

To get a full overview of the performance of the MPC, a static and a dynamic scenario are carried out in the experimental validation. Therefore, the outlined experimental setup is used. In the static scenario, the AGV stays in a static position while the robot tries to approach the truck cabin. In respect to the AGV's position, the robot starts with a relative deviation of 1610 mm, -77 mm and -74 mm for x-. y- and z-direction (in the robot world coordinate system). Fig. 4 shows that the robot reaches its target position without overshot. After 38 seconds, the robot stays on stationary position with a constant deviation of 0,53 mm, 0,17 mm and 0,13 mm for x, y and z. During the movement, a maximal robot velocity of 200 mm/s is considered concerning the velocity of the windshield gripper in the real assembly process. The velocity limit is set as an MPC constraint. The resulting control deviation satisfies the requirements of windshield assembly, which are typically specified with a tolerance of 1.5 mm. Apart from the control deviation, the robustness of the control can be seen when iGPS errors, due to lost signals, appear in the shape of steps in the course of the deviation in x-direction at 2, 5 and 13 seconds. The trajectory dynamics remain unaffected by sudden changes in signal.

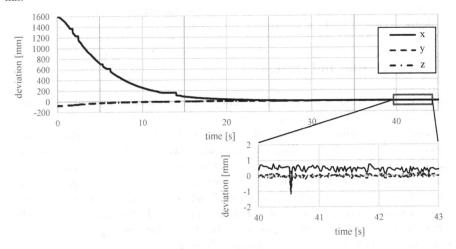

Fig. 4. Control deviation in x, y and z between truck cabin and robot for a stationary object

5.2 Tracking of a Dynamic Moving Object

In the dynamic scenario, the AGV moves with a constant velocity of 50 mm/ s on a straight line. The robot starts with a deviation to the AGV position of -460 mm, -118 mm and 7 mm for x, y and z. After activating the MPC, the robot follows the trajectory of the AGV and tries to synchronize to it. Fig. 5 illustrates the development of the control deviation, while the synchronization process takes place.

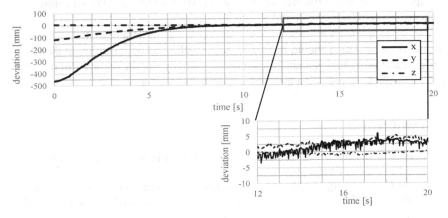

Fig. 5. Control deviation for tracking of the AGV trajectory at 50 mm/ s

Obviously, there is also no overshot and the deviation asymptotically converges to a constant offset of 3.40 mm, 3.72 mm and 0.08 mm for x, y and z. The constant deviation is slightly increased in comparison to the first test. The process velocity is decreased from 100 mm/s (in the simulation) to 50 mm/s, because higher velocities lead to bigger deviations with the current parameter setup.

6   Conclusion and Outlook

The integration of assembly processes dealing with both large components and narrow tolerances into the moving assembly line can only be realized by the use of feedback controllers. Simulation results have shown that the model predictive control achieves synchronization between handling and conveying system. The final validation was carried out on a full-scale demonstrator. As an experimental result, a maximal control deviation of 0.53 mm and 3.72 mm has been achieved for the static as well as the dynamic case. Considering these results, MPC shows feasibility to synchronize the robot to the movement of the product. For an integration in a real production environment such as a truck assembly line, further investigations must be carried out regarding the metrology system and the prediction of the target trajectory.

In future works, it has to be investigated why there is a difference between experimental and simulation results. Due to the fact, that the indoor-GPS as part of the control system was used to evaluate the dynamic behavior, a reference system such as a laser tracker should be used to detect the real behavior of the AGV and the robot. The dynamic uncertainty of 0.72 mm of the indoor-GPS is too big for achieving the required tolerances of 1.5 mm. The technical limitations of the indoor-GPS such as the measuring rate of 40 Hz and the dead time of 180 ms require wide prediction, which impede accurate position tracking of the moving truck cabin at high process speeds. In order to enhance the dynamic tracking capability of the indoor-GPS, the measuring system can be improved with an inertial measurement unit (IMU), which calculates the movements of the truck cabin between the iGPS measuring points.

The MPC approach allows numerous opportunities in parameter tuning, so that a parameter optimization should be carried out as well as an optimization of the internal model of the MPC. In this manner, the process velocity and the accuracy of the assembly can be further increased approaching industry standards.

References

1. Prasch, M.: Integration leistungsgewandelter Mitarbeiter in die variantenreiche Serienmontage. Herbert Utz Verlag, Munich (2010).
2. Storm, C., Schönberg, A.: Model Predictive Control Approach for assembling Large Components in Motion. Production Engineering - Research and Development 11(2), 167-173 (2017).
3. Reinhart, G.; Werner, J.: Flexible Automation for the Assembly in Motion. Annals of the CIRP, vol. 56, pp. 25-28 (2007).

4. Cheng, M.; Li, Y.: Controller Synthesis of Tracking and Synchronization for Multiaxis Motion System. IEEE Transactions on Control Systems Technology, vol. 22, pp. 378-386 (2011).

5. Werner, J.: Methode zur roboterbasierten förderbandsynchronen Fließbandmontage am Beispiel der Automobilindustrie. Herbert Utz, Verlag, Munich (2008).

6. Demeester, F., Dresselhaus, M.: Referenzsysteme für wandlungsfähige Produktion. AWK Aachener Wekzeugmaschinen-Kolloquium, pp. 449-477, Shaker, Aachen (2011).

7. HTWK Leipzig Homepage, http://www.msr.eit.htwk-leipzig.de/forschung/projekte/abgeschlossene-projekte/Projekt_modellbasierte_regelung, last accessed 2017/10/09.

8. Kühne, F., Lages, W.: Model Predictive Control of a Mobile Robot Using Linearization. International Conference on Mechatronics and Robotics 2004, IEEE Industrial Electronics Society: Mechatronics and Robotics, vol. 4, pp. 525-530 (2005).

9. Ellis, M., Durand, H: A tutorial review of economic model predictive control methods. Journal of Process Control 24, 1156-1178 (2014).

10. Schmitt, R., Schönberg, A.: Global referencing systems and their contribution to a versatile production. Proceedings of the 2011 International Conference on Indoor Positioning and Indoor Navigation, Guimarães, Portugal (2011).

11. Norman, A., Schönberg A.: Validation of iGPS as an external measurement system for cooperative robot positioning. The International Journal of Advanced Manufacturing Technology, vol. 64, pp. 427-446 (2013).

12. Wang, Z.; Mastrogiacomo, L.: Experimental comparison of dynamic tracking performance of iGPS and laser tracker. The International Journal of Advanced Manufacturing Technology, vol. 56, pp. 205-213 (2011).

Towards soft, adjustable, self-unfoldable and modular robots – an approach, concept and exemplary application

Robert Weidner[1,a], Tobias Meyer[1,b] and Jens P. Wulfsberg[1,c]

[1] Helmut-Schmidt-University/University of the Federal Armed Forces Hamburg, Laboratory of Manufacturing Technologies, Holstenhofweg 85, 22043 Hamburg, Germany
aRobert.Weidner@hsu-hh.de, bTobias.Meyer@hsu-hh.de,
cJens.Wulfsberg@hsu-hh.de

Abstract. Appropriate assistive systems are used due to the demographic change, increased temporal and spatial interaction or integration of human and robots as well as necessary flexibility in respect to tasks and location. Especially safety and flexibility for effective human-robot-interaction and a broad spectrum of tasks are important. Existing technologies use different kinds of mechanisms or principles for realizing safe interaction, especially e.g. soft control, optical approaches and sensor skin in combination with conventional industrial robots. In addition, the field of soft materials robotics is becoming increasingly important, with the focus to increase E-modulus of structure elements. This paper presents an approach for soft and self-unfolding modular robots on the basis of paper lamella technology which can be used as a system for human-machine interaction as well as for robots with the ability of self-unfolding and -folding. For demonstration, a realization of a self-unfolding lamella element as well as the results of first measurements are summarized.

Keywords: Soft Robotic, Compliance, Self-Unfoldable System, Modular Robotics, Mobile System, Human-Machine Interaction.

1 Introduction

The cooperation between humans and technology has changed in the recent years. It has been intensified due to, for example, the desired increase in productivity and ergonomic improvements, increased demands for quality and other aspects such as the demographic change. Two aspects are fostered especially: a) increased level of interaction and integration and b) increase mobility of automated systems.

Human-robot-interaction or integration. Due to the advancement, on the one hand, the safety fence can be dispensed within modern automation solutions [1]. Thus, coming from temporally or spatially separated work, today the work of human and technology are increasingly merged and synchronized. On the other hand, more and more human beings are supported by, e.g., wearable systems like exoskeletons [2] – so called hybrid systems. In both cases, robots must be designed in such a way, that they do not hurt people as well as technical systems that can realize natural human patterns [3].

© Springer-Verlag GmbH Deutschland, ein Teil von Springer Nature 2018
T. Schüppstuhl et al. (Hrsg.), *Tagungsband des 3. Kongresses Montage Handhabung Industrieroboter*, https://doi.org/10.1007/978-3-662-56714-2_27

Mobile systems. In addition to the diversity of occurring tasks, flexibility in respect to the location is becoming more important every day. Here, ground-based or airborne robots may be mentioned, as they have spread (also to new fields of application) in the recent years due to technological advances. The focus lies primarily on the systems mobility and less on the execution of traditional production tasks. Weight and space of handling devices are still very limited, especially in the case of airborne robots such as drones.

This paper presents a concept as well as a prototypical realization of a soft, modular and switchable robot, which is particularly suitable for mobile use – for example for frequent changes of its locations utilizing e.g. drones.

2 State of robots

In recent years, different kinds of soft robots have been developed, with respect to realize softness, compliance as well as flexibility in task and location. In the following, some approaches are summarized. It should be noted that some approaches or systems have been developed against the background of other issues.

2.1 Softness and compliance

Essentially, soft robots or robots with compliance can achieve the desired properties, especially softness and elasticity, using different approaches. Four main directions or combinations are possible:

- Soft control: Compliance integrated in control, setpoint by sensors [4].
- Soft materials: By thermal stresses, thermal as well as pressure fluctuations deformable soft materials like liquids, polymers, foams, gels, colloids, granular materials and most soft biological materials [5, 6].
- Soft structure elements: Targeted integration of flexible elements into the mechanical structure of technical systems [3], in order to increase flexibility or enable natural human patterns in the context of exoskeletal systems.
- Soft actuators: Actuators with inherent compliance. Here, typically material or geometric asymmetry are used to achieve motion in desired direction. [7, 8]

2.2 Flexibility in task and location

Flexibility in location. Generally speaking, robots can be classified to stationary and mobile robots, e.g., industrial robots usually are stationary or moving platforms are not fixed to any place. Both kinds of robots can interact with the environment depending on the configuration, e.g., gripper and sensors used. Mobile robots may be ground-based (e.g., moving platforms [9]) or airborne systems (e.g., drones [10]). New approaches for mobile systems are, for example, origami robots [11] – the basic approach can also contribute to improve mobility (keyword self-folding system).

Flexibility in task. Classic robots are programmable or controllable systems. Tasks usually have to be programmed by the user or the robot must be teached directly. Thus, the flexibility in respect to the tasks is usually low. In addition, robots must be adapted to the tasks in terms of, e.g., the load, range and degree of freedoms. In addition to standardized monolithic robots, there are also approaches for modular systems in order to configure task-adapted handling systems [12, 13].

2.3   Challenges and Objectives

Although classical industrial robots are designed as universal systems, robots are subject to different requirements. Depending on the application, the system must fulfill different requirements, for example depending to the mass of components to be handled or the kind of cooperation between human and technology. For the addressed context, a trade-off between stiffness, softness, mutability, compactness and mobility is necessary – see **Fig. 1**.

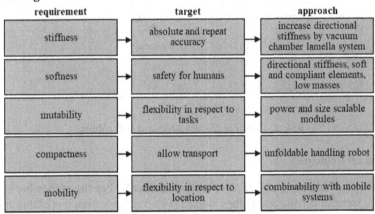

Fig. 1. Main requirements, targets for robots for addressed application and selected approaches.

3   Concept for self-foldable soft robots

The approach (and also the basic principle), which is presented in the following, describes an approach for the realization of soft robots, which also possess the ability of self-unfolding. They do not replace classic robots. Their field of application is primarily for tasks with low handling weights or low interaction and process forces.

3.1   Approach

As summarized in **Fig. 1**, the concept combines different approaches in order to realize robots for direct human-robot interaction and robots with high flexibility in respect to their location.

Fig. 2 shows the central approaches of the concept for foldable soft robots (called SFR – Soft and Foldable Robot), which can be coupled with other robots, e.g., industrial robot, mobile platforms or drones. The concept consist of two main parts: a) element(s) for realizing stiffened structure elements and b) element(s) for unfolding, extending or pushing up structure elements (also folding elements are possible). The stiffenable elements (part a) describe the classic robot segments. Thus, these are not the drives, which may also contribute to a stiffening system. The elements of part b describe the mechanism for realizing unfolding, extending or pushing up the system. Preferably, these elements are arranged parallel to the stiffenable elements. The elements must be coordinated with each other in order to realize functionality.

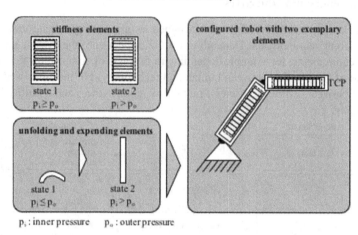

Fig. 2. Concept for foldable and soft robots.

3.2 Main Principles

Element(s) for realizing stiffened structure elements. For the stiffenable structure elements, vacuum lamella technology is used [13]. The basic principle of this approach concentrates on the use of soft chamber elements with different geometry, which are structured or non-structured and include elastic or rigid elements. By generating a vacuum inside the chambers, the elements are pressed together which strengthens the whole structure due to increased friction between the lamellae. The amount of strengthening depends, for example, to the pressure difference, kind of elements, material of elements and arrangement and interaction thereof. In principle, such elements have two, "switchable" operating states. Once the pressure inside the chamber is less than on the outside, they are active otherwise not active. Through a higher stiffness of the structure while activated, forces and torques can be transmitted.

Element(s) for unfolding, extending or pushing up structure elements. For the second functionality, the balloon effect is used. By using air pressure, elements can be raised. In addition to the differential pressure (inside the chamber and outside the chamber), the geometry and design of the element has an influence on the final shape and geometry of the stiffened element.

Element Integration

The self-unfoldable, soft structure elements – combination of the two described parts – have two airtight chambers made of thin and flexible material. Each chamber, later referred as "inner" and "outer" chamber, has a pneumatic connection. The internal pressure can be specifically adjusted by the connection (under pressure, over pressure or pressure equalization with respect to the atmosphere around). The design of self-unfoldable, soft robots depends mainly to the desired properties. Possible here is a serial and/or parallel arrangement.

In order to unfold the structure elements and to increase the stiffness of the elements, the inner and the outer chamber must be applied in a fixed sequence with over and under pressure, see **Fig. 3** as an example with cuboid geometry elements. It is assumed that both chambers as well as the chambers and the environment have the same pressure at the beginning. In the first step, both, the inner and the outer chamber are subjected to over pressure (p_1 & $p_2 > p_{out}$). The pressure between the inner (p_1) and outer chambers (p_2) must either be equalized or slightly higher in the inner chamber so that there is no pressure between the inner chamber wall and the lamellae. This would increase the stiffness of the integrated elements and thus at least complicate the process of unfolding. During this step, the unfolding process of the structural element takes place, where it takes the shape of a rod in its simplest form. In the second step, the inner chamber is evacuated ($p_1 < p_2 > p_{out}$). The reduced pressure of the inner chamber provides compression of the chamber shell by the pressure difference between the inner and outer chamber. The pressure exerted on the lamella increases the friction between the lamellae and thus ensures stiffening of the whole element. During the evacuation process, the outer chamber, which is still under pressure, maintains the form of the lamella element or the shell surrounding it. Thereafter, the system can, for example, be used for handling tasks. After performing tasks, the robot can also be folded again. In order regain the foldability of the element, the pressure of both chambers must be equalized to the pressure of the environment in order to decrease the friction between the elements.

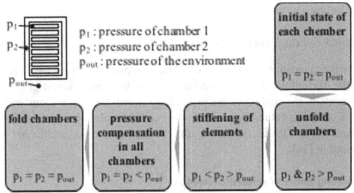

Fig. 3. Pressure-Level-Sequence for switching elements and robots with considered technology regarding chamber pressure.

4 Exemplary Realization

The design of robots based on the presented technology can be context-dependent. **Fig. 4** shows an exemplary SFR module with paper sheets on the left side (state activated) and an exemplary robot demonstrator with two active degrees of freedom on the right side.

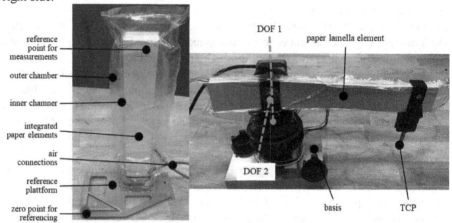

Fig. 4. Test setup for self erecting and self stiffening module based on SFR technology (left) and exemplary implementation within a hybrid robot based on ridged structural elements as well as on vacuum chamber technology (right).

5 Evaluation by Experiments

Different tests were performed to assess the introduced technology. The key results are presented below.

5.1 Objects of the Experiments

The aim of the evaluation was to review the basic approaches in respect to repeatability. In particular, the unfolding process and the holding force were investigated with exemplary elements in the context of the technology shown here. The focus was on the investigation of the repeatability accuracy during the unfolding process and the compliance under different loads. Further results for the basic vacuum chamber technology can be found in [14].

5.2 Method and Experimental Setup

To evaluate the basic approach of the main functionalities, two test cases are performed with fixed boundary conditions. The experimental setups are illustrated in **Fig. 5**. Three tests are carried out with two experimental setups:

a) Analysis of the compliance on a bending beam with setup (a), see **Fig. 5** left: Using a horizontally clamped beam element with 150 sheets of paper, geometric dimensions 297 x 45 mm in an evacuated chamber. For the investigation, the load (lever arm of 250 mm from load to clamping position) varied between 0 and 2 kg. The displacement which is caused by the load is measured by using a tape measure, measuring the position before and after applying the load.

b) Analysis absolute accuracy and repeatability accuracy during the unfolding process of one element with setup (b), see **Fig. 5** right: Starting from an initial situation (the same each time), the self-unfolding element (150 sheets of paper, geometric dimensions 297 x 45 mm) is unfolded vertical. The accuracy is examined in three spatial directions while observing the sequence described above. For unfolding, a pressure of around 100 mbar is used. The test has been repeated 10 times. The relative position of the measuring point (which is marked directly on the paper within the vacuum chamber) to the reference point is measured by using a handheld full-colour 3D-Scanner ("Spider" by Artec 3D) with a 3D accuracy of 0,03% over 100 cm. In its erected state, the setup is 3D-scanned in such a way, that the measuring point and the reference point are captured in the same scan. The collected data from the scan is then refined by using the software "Artec Studio 12", which is also used to virtually perform the final spatial measurements on the 3D-modell.

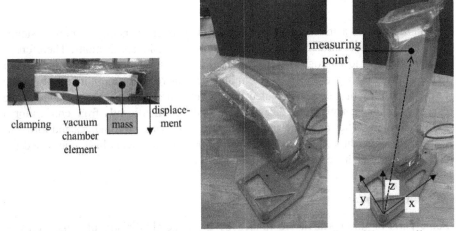

Fig. 5. Experimental setup for the evaluation: (left) for evaluation of the compliance of vacuum-chamber elements, (right) vertical self-unfolding and -stiffening of one element with reference points for 3D-measurements.

5.3 Main Results and Discussion

The experiments have shown that the basic principles a) are suitable for use as soft structures in handling systems and b) enable self-unfolding.

Compliance on a bending beam. The experiments with setup a) have shown that the compliance in vertical direction is low for masses up to 2 kg (< 2 mm). It was evident that the likelihood of twisting the element increases with increasing weight. The limit could not be determined.

Repeatability accuracy. The experiments with setup b) have shown the following accuracy:

- a mean value (MA) of 98.25 mm and a standard deviation (SD) of 1.15 in x-direction,
- an MA of 68.91 mm and SD of 7.12 mm in y-direction as well as
- an MA of 248.48 mm and SD of 0.44 mm in z-direction.

The measurements have shown good accuracies in x- and z- direction. The results in y-direction are also acceptable. The larger differences are due to the lower stiffness in this spatial direction.

General. Results from the evaluation in the laboratory show, that the described technology is generally suitable for handling objects with low weight as well as it is allowing the unfolding of structure elements. Main applications are, for example, tasks with a strong interaction with humans as well as in mobile applications with limited space – especially for handling low masses.

6 Conclusion

This paper presents a novel concept for soft and self-foldable, modular robotic systems, especially for direct interaction with humans or for mobile applications. Therefore, a vacuum chamber technology is used, which allows the individual adjustment in respect to the configuration. Moreover, this technology can increase the mobility of robots due to the possibility of a semi-automated folding and unfolding mechanism. Due to the design and the mechanisms used, such systems differ significantly from conventional industrial robots. They are primarily suitable for tasks with lower masses and forces as well as for direct human-machine interaction. First experiments have confirmed the advantages, which have been already mentioned above.

Acknowledgement

Some parts of this research are funded by the Federal Ministry of Education and Research (BMBF) in the project „smart ASSIST – Smart, AdjuStable, Soft and Intelligent Support Technologies" (founding number 16SV7114), one project in the program for an interdisciplinary build-up of competence in human machine interaction against the background of demographic changes. Supervision is provided by VDI/VDE INNOVATION GmbH. The sole responsibility for the manuscript contents lies with the authors.

References

1. Schenk, M., Elkmann, N.: Sichere Mensch-Roboter-Interaktion: Anforderungen, Voraussetzungen, Szenarien und Lösungsansätze. Demografischer Wandel – Herausforderungen für Arbeits-und Betriebsorganisation der Zukunft. GITO, Berlin, pp. 109-120, 2012.
2. Weidner, R.: Technische Unterstützungssysteme, die die Menschen wirklich wollen. Band zur zweiten Transdisziplinären Konferenz, ISBN: 978-3-86818-089-3 (print) and 978-3-86818-090-9 (online), 2016.
3. Otten, B., Weidner, R., Linnenberg, C.: Leichtgewichtige und inhärent biomechanisch kompatible Unterstützungssysteme für Tätigkeiten in und über Kopfhöhe. 2. Transdisziplinäre Konferenz "Technische Unterstützungssysteme, die die Menschen wirklich wollen", pp. 495-505, 2016.
4. Mason, M. T.: Compliance and Force Control for Computer Controlled Manipulators. In: IEEE Transactions on Systems, Man, and Cybernetics, Vol. 11, No. 6, pp. 418-432, doi: 10.1109/TSMC1981.4308708, 1981
5. Nature Homepage, https://www.nature.com/subjects/soft-materials, last accessed 2017/11/02.
6. Cheng, N. G., Lobovsky, M. B., Keating, S. J., Setapen, A. M., Gero, K. I., Hosoi, A. E., Iagnemma, K. D.: Design and Analysis of a Robust, Low-cost, Highly ArticulatedManipulator Enabled by Jamming of Granular Media. IEEE International Conference on Robotics and Automation (ICRA), Minnesota (USA), pp. 4328-4333, doi: 10.1109/ICRA.2012.6225373, 2012.
7. Li, S., Vogt, D. M., Rus, D., Wood, R. J.: Fluid-driven origami-inspired artificial muscles. Proceedings of the National Academy of Sciences of the United States of America 114 (50) 13132-13137, https://doi.org/10.1073/pnas.1713450114, 2017.
8. Drotman, D., Jadhav, S., Karimi, M., deZonia, P., Tolley, M. T.: 3D printed soft actuators for a legged robot capable of navigating unstructured terrain. IEEE International Conference on Robotics and Automation (ICRA), Singapore, pp. 5532-5538, doi: 10.1109/ICRA.2017.7989652, 2017.
9. YOUBOT STORE Homepage, http://www.youbot-store.com/, last accessed 2017/1102.
10. Kondak, K., Krieger, K., Albu-Schaeffer, A., Schwarzbach, M., Laiacker, M., Maza, I., Rodriguez-Castano, Ollero, A.: Closed-Loop Behavior of an Autonomous Helicopter Equipped with a Robotic Arm for Aerial Manipulation Tasks. International Journal of Advanced Robotic Systems, doi: 10.5772/53754, 2012.
11. Belke, C. H., Paik, J.: Mori: A Modular Origami Robot. In: IEEE/ASME Transactions on Mechatronics, vol. 22, no. 5, pp. 2153-2164, doi: 10.1109/TMECH.2017.2697310, 2017.
12. Müller, R., Esser, M., Vette, M.: Reconfigurable handling systems as an enabler for large components in mass customized production. In: Journal of Intelligent Manufacturing, vol. 24, issue 5, pp. 977-990, https://doi.org/10.1007/s10845-012-0624-y, 2013.
13. IGUS Homepage, http://www.igus.de/wpck/18353/roboticcomponents?gclid=CjwKCAjwhOvPBRBxEiwAx2nhLkVNJRzNgntWlNT2TvwPbfZUbB9D-l96wpufJNtS_OgYoX-kvn_r_aRoCW2cQAvD_BwE, last accessed 2017/11/02.
14. Weidner, R., Meyer, T., Argubi-Wollesen, A., Wulfsberg, J. P.: Towards a modular and wearable support system for industrial production. Applied Mechanics & Materials, vol. 840, pp. 123 - 131, 2016.

Improving Path Accuracy in Varying Pick and Place Processes by means of Trajectory Optimization using Iterative Learning Control

Daniel Kaczor[1], Tobias Recker[1], Svenja Tappe[1] and Tobias Ortmaier[1]

Gottfried Wilhelm Leibniz Universität Hannover, Institute of Mechatronic Systems
Appelstraße 11a, 30176 Hannover
daniel.kaczor@imes.uni-hannover.de,
WWW home page: http://www.imes.uni-hannover.de

Abstract. This paper presents a universal method for improving the path accuracy in varying, highly dynamic pick and place processes. The proposed method is based on an iterative learning control (ILC) and is valid for serial as well as parallel robots. It extends the traditional ILC approach for purely repetitive tasks by evaluating similarities between the occurring movements. Based on this, the correction term for following trajectories can be calculated and a significant improvement in the accuracy can be obtained even for previously unknown (unlearned) motions. The performance of the method is studied and verified using an exemplary four degrees of freedom delta robot. It is shown that the presented approach improves the path accuracy up to 86% even if the occurring pick and place trajectories vary with respect to start and end position. It also outperforms conventional computed torque control methods by up to 50%.

Keywords: Iterative Learning Control, Path Accuracy, Pick and Place, Machine-Learning

1 Introduction

Repetitive processes are one of the most common applications in automation technology and robotics. Due to the widespread use of robots in modern production, even the smallest increases in accuracy or speed of the motion sequences lead to a significant rise in productivity. As a result, various efforts are being made to enhance the performance of industrial robots. A possible approach to reduce the path error and vibrations is dynamics based motion planning as presented in [1]. Here, the path and the motion profile are optimized for time-optimal and smooth high speed pick and place applications based on the robot's dynamics. Another common approach is the feedforward control of the joint speed or joint torque (TFFC). More advanced feed forward control approaches are e.g. presented in [2]. In this paper, flatness-based control and adaptive input shaping are used to reduce the vibration of parallel robots during rapid movements. Precondition for all these listed methods is an appropriate model of the

© Springer-Verlag GmbH Deutschland, ein Teil von Springer Nature 2018
T. Schüppstuhl et al. (Hrsg.), *Tagungsband des 3. Kongresses Montage Handhabung Industrieroboter*, https://doi.org/10.1007/978-3-662-56714-2_28

robot dynamic. The modeling and identification of the necessary model parameters are not trivial and require expert knowledge [3], [4]. Looking at the ongoing advancement of robotics in medium-sized enterprises, it can not be guaranteed that these companies have the necessary expertise available. One of the resulting challenges is how to meet the need for more efficient production under the constraint of limited knowledge. A method which does not require knowledge of the dynamics is Iterative Learning Control (ILC) [5]. It is a learning-based pre-control that iteratively generates a correction for the manipulated variable based on the error traces of the preceding runs [6], [7]. However, a conventional ILC is limited to a fixed motion sequence. For varying tasks the performance may severely decrease [8]. Therefore, in this paper, a conventional ILC is extended by a function to utilize the similarities of occurring trajectories when varying either the pick or the place position or both. With this modification the ILC is allowed to control a class of similar movements by approximating unknown trajectories by known trajectories with the help of regression. This self-learning method for increasing the path accuracy of robots in repetitive processes, which is also applicable to varying pick and place tasks, is presented in this paper. The potential of this method is demonstrated for three different complexity levels of pick and place applications. Furthermore, the methodology is compared for a general task with a classical TFFC regarding the criteria of the achieved path accuracy and settling time.

This paper is organized as follows: Section 2 describes the four scenarios of pick and place applications examined in this paper. Section 3 gives a short view on the theoretical and mathematical background of the adapted iterative learning control method. Experimental setup and experimental results are given in Section 4, demonstrating the high potential of the proposed method. Moreover, a comparison to a state-of-the-art TFFC [2] is presented. Section 5 closes this paper with the conclusion and outlook on future work.

2 Pick And Place Scenarios

This paper differentiates between two main categories and four different scenarios of a pick and place applications, see Fig. 1. The first category includes purely repetitive processes. This means that the pick as well as the place position are invariant over the entire process. The robot always moves exactly the same way (Scenario I). The second category describes a set of variable processes within a typical handling task. In this paper, the assumption is made that these may vary but still have a substantial similarity regarding the geometric path and are executed inside the defined workspace. Within this category, there are three more possible scenarios. The first describes processes in which the pick position of the object to be manipulated is invariant, but the place position can vary between discrete known positions (Scenario II). Scenario III describes processes, in which the pick position is variable and the place position can also vary between discrete known positions. The third and most complex application is shown in

Scenario IV. Here, both the pick and the place positions are variable within the predefined working area.

Within this paper, we will discuss only the Scenarios III and IV since these are the most challenging to the developed methodology.

Fig. 1. Examined pick and place scenarios (I-IV), with pick positions within the orange framed area (left) and place positions within the blue frame (right)

3 Iterative Learning Control for Varying Processes

As described above, learning controllers are optimized to be used in invariant processes. Let us assume a robot being supposed to move a object from one position to another. The robot performs a standard pick and place motion consisting of two horizontal sections and an elliptical section in between. Due to various effects, e.g. inertia or friction, a deviation between nominal and actual path is caused. An example of this deviation is shown in Fig. 2 (a). Based on this error, the ILC calculates a correction term c_j for the system input x_j of the subsequent iteration k, see Fig. 2 (b) [5]. This term is adapted iteratively in each cycle so that the error is minimized step-wise (Fig. 2 (a)). In this case, the reference path is used as a criterion for the ILC in contrast to the system input typically used in the literature. This is necessary to build a universal method and to optimize the path accuracy even in the case, where the access to the axis control is limited. Once the path error falls below a defined threshold, the ILC is completed, the current correction is saved and applied to any further trajectories. The approach used in this paper is to transfer this correction from one learned path j with $1 \leq j \leq n$ to an unlearned path i with $i = n+1$, assuming n equals the total number of previously learned path. For this purpose, several distinctive and equally distributed paths in the application area are learned using a conventional ILC. For the calculation of the pre-control term for a new path

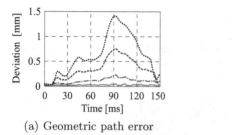

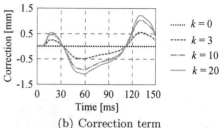

(a) Geometric path error (b) Correction term

Fig. 2. Development of the correction term (a) and geometric path error (b) over 20 iterations

i, the similarity $\sigma_{i,j}$ (Eq. 1) to the previously learned paths j is determined. For this purpose, the quadratic difference to the corresponding point on the new target path x_i is calculated for each sample s of the learned paths n (N = total number of samples).

$$\sigma_{i,j} = \sqrt{\sum_{s=1}^{N} (x_i(s) - x_j(s))^2} \quad (1) \qquad \sigma_i = [\sigma_{i,1}, \dots, \sigma_{i,n}]^{\mathrm{T}} \quad (2)$$

$$g_{i,j} = 1 - \frac{\sigma_{i,j}}{\max(\sigma_i)} \quad (3)$$

To calculate the weighting $g_{i,j}$ of each individual correction term c_j, the similarity $\sigma_{i,j}$ to the corresponding path x_i is divided by the maximum of all calculated similarities, see Eq. 2 & Eq. 3. A high degree of similarity results in a high weighting of the corresponding correction term. To correct an unknown trajectory, each known correction term is multiplied by the corresponding weight (Eq. 4). The resulting correction term c_i is then added to the path profile r_i, see Eq. 5.

$$c_i = \sum_{j=1}^{n} g_{i,j} c_j \quad (4) \qquad r_i^* = r_i + c_i \quad (5)$$

4   Results

The effectiveness of the proposed methodology is studied and verified using a four degree of freedom (4-dof) delta robot. Following the short introduction of the experimental setup in Section 4.1, the experimental results are presented in Section 4.2.

4.1 Experimental Setup

For experimental evaluations and demonstration of the potential the presented approach was tested on a 4-dof delta parallel robot and a serial 4-dof SCARA manipulator (Fig. 3) of the model factory, which was presented in [9]. In addition to the above-mentioned mechanisms, the plant also contains a 3-dof small-scale stacker crane, and two conveyor belts. In this paper only the results of the experiments on the delta robot are presented. However, the shown results are also transferable to the SCARA manipulator as the ILC algorithm is not restricted to a specific hardware. The used kinematic is an industrial Codian D4-1100 delta robot designed for high dynamic pick and place applications and it is controlled by a standard industrial PLC and servo inverters.

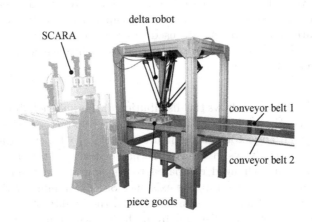

Fig. 3. Experimental setup consisting of a 4-dof delta robot, a 4-dof SCARA manipulator, and two conveyor belts

4.2 Experimental Results

The capability of the presented approach is demonstrated based on two slightly different examples. The first example involves a pick and place process in which parts are picked from a random position P_{S_i} and placed at a random position P_{E_i}. The point P_{S_i} varies in a defined range, whereas the point P_{E_i} is restricted to a limited number of nine discrete points.

Table 1. Possible pick and place positions depending on the respective scenario

Position	Scenario III	Scenario IV
$\boldsymbol{P}_{S_i}$	$\begin{pmatrix} x_{S_{\min}} \leq x_{S_i} \leq x_{S_{\min}} \\ y_{S_{\min}} \leq y_{S_i} \leq y_{S_{\max}} \\ z_{S_i} \end{pmatrix}$	$\begin{pmatrix} x_{S_{\min}} \leq x_{S_i} \leq x_{S_{\min}} \\ y_{S_{\min}} \leq y_{S_i} \leq y_{S_{\max}} \\ z_{S_i} \end{pmatrix}$
$\boldsymbol{P}_{E_i}$	$[\boldsymbol{P}_{E_1}, \boldsymbol{P}_{E_2}, ..., \boldsymbol{P}_{E_9}]$	$\begin{pmatrix} x_{E_{\min}} \leq x_{E_i} \leq x_{E_{\max}} \\ y_{E_{\min}} \leq y_{E_i} \leq y_{E_{\max}} \\ z_{E,i} \end{pmatrix}$

This scenario is the equivalent of picking a component from a moving conveyor belt (which leads to a variable pick position) and placing it in a device or package (Scenario III). The second test scenario differs in that the place position $\boldsymbol{P}_{E_i}$ is also variable now (Scenario IV), see Table 1. To provide meaningful information about the accuracy improvement a number of 600 trajectories between random start and end points are used. Each trajectory is measured with and without the ILC. The path error occurring in each movement is recorded and the improvement is determined.

Scenario III Figure 4 shows the distribution of the summed absolute path error over the area of the application. The x- and y-axes characterize the starting point of the particular trajectory. In the figures (4 (a) & (b)) shown, the end point of the trajectories is always identical. On the vertical axis the factor with which the path error has changed compared to the initial state (no pre-control) is specified. The figures illustrate the existence of local minima at the positions of the grid points used for regression. Generally, the quality of the approximation decreases with increasing distance from the known points. This is caused by the greater uncertainty in the interpolation between known trajectories of the required correction term. The values shown in Table 2 are taken from a series of measurements with 600 random trajectories. On average, the error is reduced by 86% compared to the initial state. In general, a higher number of initial grid points or a smaller distance between them leads to a higher overall accuracy, see Fig. 4 (b).

Table 2. Scenario III: Improvement of controller performance by using the ILC

Pre-control	Average error	Max. error	Overshoot	Settling time
none	100%	117%	1.38 mm	12 ms
ILC (4 grid points)	19.1%	23.4%	0.34 mm	7 ms
ILC (9 grid points)	14%	18.4%	0.23 mm	5 ms

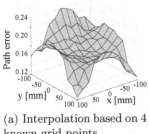

(a) Interpolation based on 4 known grid points

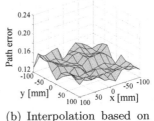

(b) Interpolation based on 9 known grid points

Fig. 4. Variation of the accumulated path error depending on the position in the workspace

Scenario IV This section is based on an application in which both the pick as well as the place positions can be located anywhere in the previously defined workspace. The considered scenario represents a much greater challenge for the algorithm since the similarity of the occurring trajectories is significantly reduced. In order to compensate for this, the ILC is trained with 200 and 600 random trajectories. As the experimental results in Table 3 show, it is necessary

Table 3. Scenario IV: Improvement of controller performance by using the ILC

Pre-control	Average error	Max. error	Overshoot	Settling time
none	100%	116%	1.38 mm	12 ms
ILC	53.7%	90.6%	1.08 mm	10 ms
ILC trained 200	28.4%	65.3%	0.78 mm	8 ms
ILC trained 600	18.3%	28.5%	0.34 mm	7 ms

to learn a larger number of correction terms due to the limited similarity of the occurring trajectories. It can be shown that with the help of prior training, the average error can be significantly reduced. However, the accuracy of the ILC is generally lower than in the previously considered scenarios, despite the learning process. This shows the main disadvantage of the proposed approach. The reliance on the similarity of the occurring trajectories represents the greatest limitation of the proposed approach.

Comparison with TFFC In order to classify the results more accurately, the ILC is compared with a TFFC. Below, both methods are compared using the scenarios introduced in Section 4. Figure 5 shows the distribution of the cumulative path error of 600 random pick and place trajectories for Scenarios III and IV. It is important to note that the criterion used is not the following error, but the actual geometric path error. The path error does not contain a time component and, therefore, penalizes deviations in the path more severely than the tracking error. For an application as described in Section 4.2, the ILC

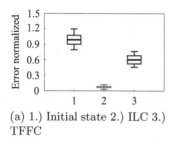

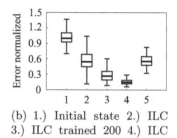

(a) 1.) Initial state 2.) ILC 3.) TFFC

(b) 1.) Initial state 2.) ILC 3.) ILC trained 200 4.) ILC trained 600 5.) TFFC

Fig. 5. Distribution of the cumulative path error normalized to the average initial value ((a): Scenario III; (b): Scenario IV)

achieves an average path error three times lower than the TFFC. Therefore, the use of an ILC is recommended for a task with a similar movement pattern. In an application with varying pick and place points, there may be greater variation in accuracy. Nevertheless, the ILC is on average twice as accurate with a suitable number of learned correction terms.

5 Conclusion and Future Work

This paper illustrates a universal method to improve the path accuracy for general serial and parallel robots. The method is based on an iterative learning control and extends the standard method for purely repetitive processes by the possibility to improve varying processes based on the similarity of the occurring robot motions. This allows the iterative improvement of the path accuracy even for varying pick and place processes. It was shown, that the presented method can achieve an improvement in accuracy up to 86% and in the settling time about 40% - 50% compared to the initial setup. Furthermore, the presented method is up to 50% better than a standard TFFC, depending on the considered scenario. An equally important point besides improving accuracy is the fact that the configuration of the presented algorithm is largely automatic. No qualified personnel is required for setup. Furthermore, the learning properties of the ILC allow for dynamic adaptation to changing process parameters, e.g. due to mechanical wear.

Future work will focus on the experimental validation on several kinematics to further confirm the presented method. An additional extension is the reduction of the computing effort by decreasing the size of the necessary data by calculating the correction term using a spline function with a discrete number of intermediate points instead of using all data points during the robot motion. Moreover, the implementation of the presented approach in a parameter management concept as shown in [9] and [10] to connect the learned information to other robots within a factory would be effective.

Acknowledgment The authors like to thank Lenze SE for providing the control cabinet (including power electronics components and PLCs) and for the support during this research.

References

1. Zhang, Y., Huang, R., Lou, Y., Li, Z.: Dynamics based Time-Optimal Smooth Motion Planning for the Delta Robot. 2012 IEEE International Conference on Robotics and Biomimetics (ROBIO)
2. Öltjen, J., Kotlarski, J., Ortmaier, T. On the Reduction of Vibration of Parallel Robots using Flatness-based Control and Adaptive Input Shaping, 2016 IEEE International Conference on Advanced Intelligent Mechatronics
3. Do Thanh, T., Kotlarski, J., Heimann, B., Ortmaier, T.: Dynamics identification of kinematically redundant parallel robots using the direct search method. Mechanism and Machine Theory, Volume 55, September 2012, Pages 104-121
4. Isermann, R., Münchhof, M., Identification of Dynamic Systems. Springer-Verlag Berlin Heidelberg, 2011
5. Moore, K. L.: Iterative Learning Control for Deterministic Systems. Springer, London 1993.
6. Bristow, D. A., Tharayil, M, Alleyne, A. G.: A survay of iterative learning control - A learning-based method for high-performance tracking control. Control Systems, IEEE, vol. 26, no. 3, pp. 96-114, 2006
7. Rogers, E., Galkowski, K., Owens, D. H.: Control Systems Theory an Applications for Linear Repetitive Processes. Springer, vol. 349, 2007 .
8. Gunnarsson, S., Norrlöf, M.: On the disturbance properties of high order iterative learning control algorithms. Automativa, vol. 42, no. 11 pp. 2031−2034, 2006
9. Öltjen, J., Beckmann, D., Hansen, C., Maurer, I., Kotlarski, J., Ortmaier, T.:Integrated Parameter Management Concept for Simplified Implementation of Control, Motion Planning, and Process Optimization Methods. Applied Mechanics & Materials Vol. 840, p114-122, (2016) Trans Tech Publications, Schweiz
10. Maurer, i., Riva, M., Hansen, C., Ortmaier, T.: Cloud-based Plant Process Monitoring on a Modular and Scalable Data Analytics Infrastructure. In: Schüppstuhl T., Franke J., Tracht K. (eds) Tagungsband des 2. Kongresses Montage Handhabung Industrieroboter. Springer Vieweg, Berlin, Heidelberg

Path Guiding Support for a Semi-automatic System for Scarfing of CFRP Structures

Rebecca Rodeck[1] and Thorsten Schüppstuhl[1]

[1] Hamburg University of Technology, Institute of Aircraft Production Technology, Denickestr. 17, 21073 Hamburg, Germany
rebecca.rodeck@tuhh.de

Abstract. In this paper, different approaches for the path guidance of a support system for scarfing of CFRP structures are presented. First, the machine concept, mechanical setup, and the control system are introduced. The system addressed is a kind of support system where the human operator is responsible for planar movement while an axis perpendicular to this plane is controlled automatically depending on the planar position. In the following chapter, concepts and findings for path guiding, i.e. the manual guidance of the tool on a specified path, are presented. Three basic questions have to be addressed: an appropriate execution strategy has to be found that determines on what kind of path the operator should guide the system, the user guidance along a predefined path deals with the question of how to guide the user along this path, and in the last subsection the visualization and possible use of augmented reality are discussed.

Keywords: Support System, Human-Machine-Interface, Scarfing of CFRP.

1 Introduction

Carbon fiber reinforced polymers (CFRP) are becoming more and more widespread, especially as a material with superior characteristics for the aviation industry. With the Airbus A350 and the Boeing 787, composite repair becomes increasingly important because these commercial aircrafts' structures are made more than half of composites and even their fuselages are comprised of CFRP.

The prevalent repair technique for conventional aircraft with structures made of metal is doubler repair, where a metal sheet is fastened on top of the damage using rivets. However, this kind of repair is not appropriate for CFRP due to stress concentrations occurring around the bolt holes [1, 2]. A process that is far more appropriate is the scarf repair, where the damaged as well as the surrounding healthy material is removed in the form of a scarfing with a defined scarfing ratio [3]. Subsequently, a matching CFRP patch can be bonded to the prepared surface [4] (see Fig. 1).

Today, scarfings usually have circular contours. Because these require a lot of healthy material to be removed, there has been some research on optimal shapes. It was found that the optimal shape in terms of loading are nearly elliptical [3].

© Springer-Verlag GmbH Deutschland, ein Teil von Springer Nature 2018
T. Schüppstuhl et al. (Hrsg.), *Tagungsband des 3. Kongresses Montage Handhabung Industrieroboter*, https://doi.org/10.1007/978-3-662-56714-2_29

Most of the time, the scarfing process is carried out manually. However, a major drawback is the difficulty in the achievement of the required geometrical accuracy, particularly in depth, even for especially qualified and very experienced workers. Other problems are the lack of repeatability and documentation as well as the long duration of the process that can take from several hours to days [5, 6]. Hence, various efforts have been undertaken to automate the scarfing process, e.g. the STARC system [6], ULTRASONIC mobileBLOCK [7], and the system developed in the CAIRE project [8-10]. While some of these systems are functional and can be used for a wide variety of repair cases, they remain to be too heavy to be transported and secured by a single worker without any additional means such as mobile transport systems or ceiling cranes. Furthermore, the investment costs for such systems are rather high. In order to facilitate a high quality repair in simple repair cases (small defects on slightly curved surfaces), a light-weight, low-cost support system is needed that can help even inexperienced workers to achieve the desired accuracy with a significantly reduced expenditure of time.

Fig. 1. Removal of damage by scarfing and subsequent bonding of patch

2   System Setup

Since the setup of the developed system is in many ways different from most known systems, the general machine concept as well as the mechanical setup and the control system are described in the following section.

2.1   Machine Concept

A promising approach from medical engineering is known as Navigated Control (NC). Here, the position of a surgical instrument is tracked continuously to determine the distance to preoperatively defined workspace borders. When reaching these borders, the instrument is stopped automatically to avoid the injury of risk structures [11].

A similar approach can be applied to the scarfing of CFPR structures. However, switching a milling spindle on and off depending on the position in 3D space is not an option because high frequency switching is not possible due to inertia and also not suitable for conventional milling spindles. Controlling the cutting speed is not possible either because of process requirements and would not be useful in any case. Instead, in order to mill a three-dimensional contour with high accuracy, it would be appropriate to control the infeed (movement perpendicular to the area to be machined). However, letting the user guide a tool freely in 3D space and controlling the infeed with an additional, powered axis would lead to too much freedom in movement that would put additional strain on the user while trying to move the tool in a favorable way (see [12]). Further, a suitable position tracking system such as a laser scanner is expensive and would likely lead to problems with occlusion. Thus, a system is developed that restricts user movement to a defined set of axes, measures the movement in the unrestricted directions and controls the infeed dependent on the position.

2.2 Mechanical Setup

While at least five axes are required to move the milling tool freely in three-dimensional space, in this case, in order to save machine weight, the orientation of the tool cannot be adjusted. This simplification is possible because the system is to be used only on slightly curved surfaces. These considerations lead to a setup consisting of two axes for the manual movement on a plane roughly parallel to the surface to be machined and an additional, powered axis for the infeed (Fig. 2).

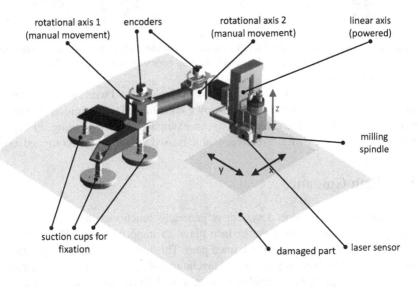

Fig. 2. Concept of articulated arm on slightly curved surface: the movement in x- and y-directions (blue plane) is done manually while the movement in z-direction is controlled automatically depending on the x, y-position

Several different kinds of kinematic configurations were investigated and assessed by the following criteria: weight, number of parts, machining at workspace boundaries, transportability, and suitability for manual movement (for a detailed description refer to [13]). An articulated arm configuration was chosen mostly because of advantages in weight, easy transportability by "folding" it to a small pack size, the possibility of using the entire workspace for machining unobstructed by the machine structure, and especially for its suitability for manual movement. The articulated arm with its rotational axes is more favorable than other configurations using linear axes due to their susceptibility to tilting. Also, in contrast to some other configurations, the articulated arm offers a mostly unobstructed view of the workspace. A corresponding mechanical system was designed and assembled.

2.3 Control System

The scarfing process consists of four general steps (Fig. 3): first, the system is positioned and fixed to the surface using three vacuum cups. Then, because the geometry of the damaged surface is not known beforehand, the surface is digitized using a laser sensor mounted to the mechanical system (Fig. 2) that is moved over the surface manually. In the third step, the desired scarfing geometry is defined based on the acquired data. Finally, the scarfing is milled by the user guiding the tool over the surface while the infeed is controlled automatically.

Fig. 3. Scarfing process consisting of four steps: fixation with vacuum cups, digitalization of damaged surface, definition of scarfing geometry, and scarf machining

The machine control was implemented in Beckhoff's TwinCAT 3 real time environment that runs on a standard windows notebook. The definition of the desired scarfing geometry is carried out using an external software running on the control notebook. During scarfing, a reference table is used to control the motorized axis [13].

3 Path Guidance

Even though the introduced system is generally functional, the path on which the milling tool is moved over the surface plays an important role. The user has to be instructed to guide it along a predefined path. This task can be divided into three main areas: the general type of path during machining of the scarfing is determined by the execution strategy, the user guidance along a predefined path helps the user move the tool along a corresponding path, and in the last subsection, the advantages of the use of augmented reality (AR) for visualization are presented.

3.1 Execution Strategy

In order to determine the general type of path for machining, different factors have to be taken into account. For the machine, a spiral or elliptical path (Fig. 4 left) would be favorable because on this type of path, the changes in infeed are minimal. On the other hand, it is assumed that the user would have difficulties following such a path. From the user's point of view, an approximately linear path might be easier to follow, though for the machine, this type of path leads to the greatest changes in infeed and is thus not optimal. Also, the requirements of the milling process have to be considered.

While the requirements of the milling process (mostly feed rate) can be taken from literature and the limitations of the mechanical system can be identified by experiments, studies have to be conducted to determine which kinds of paths are easy for the user to follow. In a first study, different kinds of linear and circular paths are compared. Several influencing factors were taken into account for the design of this study: on the one hand, factors relating to the mechanical configuration that can be identified beforehand like the change of joint angles during the manual movement along a path and the degree to which the two rotary joints are involved in the respective movement. On the other hand, there are factors that relate to the user and that are dependent on his body's geometry like the movement of the human arm while the system is guided along a given path and the position of the human in relation to the mechanical system. Due to the dependence on the individual test person, it is not always possible to investigate these factors separately.

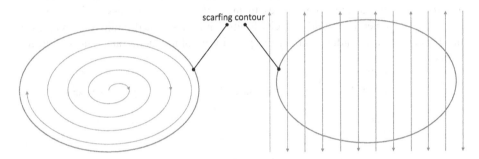

Fig. 4. Different types of paths during machining: spiral/elliptical (left) and linear (right)

The test persons are first asked to move the system in a manner that seems easy to them. Subsequently, they follow some given circular and linear paths. Finally, they are again asked for their preferred type of path.

Although the study is not yet completed, preliminary findings show a preference for circular paths. They seem to be more intuitive and easy to execute, at least when using an articulated arm mechanical system. For linear paths, although the assessment differed between the various test persons, it shows that whether a linear path is easier or harder to follow than a circular path depends on the kind of linear path and the above mentioned influencing factors. Because during scarfing, the full path consists of several linear or circular paths in order to machine the entire scarfing area, a linear execution strategy would result in a path consisting of single paths with varying difficulties. With circular paths, this variance is significantly smaller because the effect of the influencing factors is minimal, making circular or elliptical paths easier to follow.

In the next step, this will have to be evaluated in presence of process forces. It is expected that this will lead to a natural guidance, especially when using circular paths because in this case, the process forces will be approximately constant during the respective circumventions. When following linear paths, due to the changes in infeed, the changes in process forces will be large, thus making the path harder to follow. A study has to be conducted to verify these presumptions.

3.2   User Guidance along a Predefined Path

Once the general type of path for machining is known, the user has to be guided along that path. From the preliminary findings in relation to the execution strategy, a circular or elliptical path can be assumed. The full path needs to cover the entire scarfing area and can be divided into several elliptical paths with different half-axes. During machining, appropriate half-axes are chosen. The user is then guided along the corresponding elliptical path. After a full circumvention the half-axes are adapted and the user is guided along the new ellipse.

In addition to the guidance of the position along the path, the velocity also plays a role. From the machine's point of view, it is important that the tool is not moved too quickly because the controller for the infeed axis can only react to user movement, thus the error will be bigger for faster movement. The milling process itself restricts the velocity. Within the given limits, the user may choose the velocity of movement according to his personal preferences to make the process as easy and intuitive as possible. The adherence to the velocity limits can be visualized intuitively and mostly independent of the guidance along the path by a color scheme analogical to traffic lights.

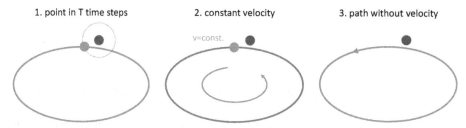

Fig. 5. Three different approaches to user guidance. Blue: actual position, green: reference position/path

Three different approaches for user guidance along a predefined path are developed (Fig. 5). Although it is not imperative, some of these approaches use reference velocities to facilitate guidance.

The first approach uses a point that is displayed to the user to indicate desired future movement. It marks the tool's desired position in a predefined number of time steps. For this, a velocity comprised of the current velocity and possibly a reference velocity is used to extrapolate the current position. The corresponding point on the reference path will be displayed to the user. To facilitate the anticipation of desired future movement, the elliptical path should also be visualized.

The second approach guides the user by rotating a point along the elliptical path with a constant reference velocity largely without consideration of the actual user movement. A further visualization of the reference path will likely not be necessary since the constant velocity makes it easy to anticipate the point's movement.

No velocity guidance is given with the third approach. Only the reference path and the direction in which to guide the tool along the path are visualized.

All of these approaches have advantages and disadvantages (see Table 1). Approach 1 responds to user movement, the influence of the reference velocity can be adjusted, and the desired movement is always apparent to the user. However, the visualized reference point can never be reached, even when the user follows the path perfectly, which might lead to demotivation. Additionally, because the entire elliptical path is visualized, the degree of obstruction of the user's field of view is high. In contrast to this, in approach 2, the reference point can be reached, leading to a direct feedback of the user's success, and the future route is easy to anticipate even without visualizing the elliptical path. The desired movement along the reference ellipse is apparent. The downsides of this approach are that it does not respond to the user's movement and that a reference velocity has to be set although technically, this is not necessary. This leads the user to guide the tool corresponding to a velocity that likely differs from his preferred velocity. The third approach does not specify a velocity. There is no response to the user's movement but with this approach, this is not a problem because the user can see the entire desired elliptical path and follow it using his preferred velocity. Similar to approach 1, the user's field of view is obstructed. However, the desired movement is not always obvious, especially when the deviation from the reference path is large. Then, it is not apparent to the user on what path to move back to the reference ellipse. The user can see how close he is to the reference path, thus giving him some feedback of success but due to the non-existence of a path leading him back after deviations it is not clear to the user whether his chosen path is a good option.

Table 1. Assessment of the different approaches to user guidance

	Approach 1: point in T time steps	Approach 2: constant velocity	Approach 3: path without velocity
Response to user movement	+	-	o
Influence on user's velocity	+	-	+
Feedback of user's success	-	+	o
Degree of obstruction of user's field of view	-	+	-
Clarity of desired movement	+	+	-

The assessment does not lead to a clear favorite, nor does it specifically preclude any approaches. They were implemented and tested, but a study will have to be conducted in order to determine which approach is favored by a representative group of users.

3.3   Visualization and Use of Augmented Reality

In order to guide the user along a path, an approach to user guidance (see Section 3.2) has to be chosen and visualized. A conventional way to do so is the use of a display such as a tablet computer (Fig. 6 left). The problem with this type of visualization is that it requires the user to take his gaze away from the actual machining area to look at the display, even when the display is fixed close to the machining area. This is not desired since the user is ought to observe the milling process. The constant disruption of having to switch his gaze between the milling area and a display is assumed to put additional strain on him. To overcome these problems, the use of augmented reality (AR) is a promising approach, implementing the visualization on a head-mounted display (HMD) such as the Microsoft HoloLens or, alternatively, by using a projector.

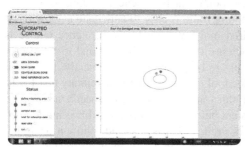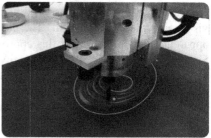

Fig. 6. Visualization of scarfing process with inner and outer contour as well as reference point: conventional display (left) and AR (right)

Besides supporting the user during milling, the use of AR is also an attractive option to add support during other process steps, e.g. the fixation of the mechanical system at an optimal position on the workpiece (aircraft) and the digitalization by visualizing the progress.

A basic visualization using AR was implemented on the Microsoft HoloLens. Overlaying the path guidance information with the actual milling area has shown to be an enormous facilitation during first experiments. This, too, will have to be validated in a future study.

4 Conclusion

The way in which the user manually guides the milling tool over the surface has a notable effect on the result when using a semi-automated system as introduced in this paper. Thus, approaches for path guidance are introduced. Considerations for the optimal type of path are shown and the preliminary results of a study suggest a preference for circular or elliptical paths. These results will have to be verified with a larger number of participants. Also, the results will have to be evaluated under the influence of process forces in another study.

Different approaches for the user guidance along a predetermined path are introduced and assessed. The assessment does not show a clear favorite, neither does it specifically preclude any approaches. A study will have to be conducted in order to determine the approach best suited.

Finally, the advantages of the use of augmented reality are pointed out. The complete implementation of the visualization using AR is expected to help facilitate the following of a given path and the semi-automatic milling process in general. This will have to be verified in another study.

Acknowledgements. This project was funded by the German Federal Ministry for Economic Affairs and Energy (BMWi, ZIM).

References

1. Ataş, A., Soutis C.: Subcritical damage mechanisms of bolted joints in CFRP composite laminates. Composites Part B: Engineering 54, 20-27 (2013).
2. Camanho, P. P., Matthews, F. L.: A Progressive Damage Model for Mechanically Fastened Joints in Composite Laminates. Journal of Composite Materials 33(24), 2248-2280 (1999).
3. Wang, C. H., Gunnion, A. J.: Optimum shapes of scarf repairs. Composites Part A: Applied Science and Manufacturing 40(9), 1407-1418 (2009).
4. Wachinger, G., Thum, C., Scheid, P.: Reparaturfähigkeit und Reparaturkonzepte bei Strukturen aus faserverstärkten Kunststoffen (FVK). In: Handbuch Leichtbau: Methoden, Werkstoffe, Fertigung, pp. 1161-1188. Carl Hanser Verlag, München (2011)
5. Holzhüter, D., Pototzky, A., Hühne, C., Sinapius, M.: Automated Scarfing Process for Bonded Composite Repairs. In: Adaptive, tolerant and efficient composite structures, pp. 297-307. Springer Berlin Heidelberg, Berlin Heidelberg (2013).
6. Erlbacher, E., Godwin, L.: Automated Scarfing and Surface Finishing Apparatus for Complex Contour Composite Structures. PushCorp Inc., Chicago (2000).
7. DMG MORI: ULTRASONIC mobileBLOCK // ULTRASONIC 85 / 260 / 360, https://en.dmgmori.com/news-and-media/technical-press-news/news/ultrasonic-mobileblock----ultrasonic-85---260---360, last accessed 2017/09/18
8. Höfener, M., Schueppstuhl, T.: Small industrial robots for on-aircraft repair of composite structures. In: Proceedings for the joint conference of 45th International Symposium on Robotics and 8th German Conference on Robotics. VDE Verlag GmbH, Berling (2014).
9. Höfener, M., Schüppstuhl, T.: A method for increasing the accuracy of "on-workpiece" machining with small industrial robots for composite repair. Production Engineering – Research and development 8(6), 701-709 (2014).
10. Höfener, M., Thum, C., Schüppstuhl, T.: Roboter zur mobilen spanenden Bearbeitung von CFK-Strukturen. VDI Z-Integrierte Produktion 10, 28-30 (2012).
11. Luz, M., Manzey, D., Mueller, S., Dietz, A., Meixensberger, J., Strauss, G.: Impact of navigated-control assistance on performance, workload and situation awareness of experienced surgeons performing a simulated mastoidectomy. The International Journal of Medical Robotics and Computer Assisted Surgery 10(2), 187-195 (2014).
12. Weidner, R., Rodeck, R., Wulfsberg, J., Schüppstuhl, T.: Unterstützung manueller Tätigkeiten: Am Beispiel des qualitätskritischen Prozesses des Schäftens von CFK-Strukturen. wt-online 9, 624-630 (2016).
13. Rodeck, R., Schüppstuhl, T.: Repair of Composite Structures with a Novel Human-Machine System. In: Proceedings of the 47th International Symposium on Robotics, pp. 660-666 (2016).

Printed in the United States
By Bookmasters